COLLISIONS OF ELECTRONS WITH ATOMS AND MOLECULES

PHYSICS OF ATOMS AND MOLECULES

Series Editors

P. G. Burke, *The Queen's University of Belfast, Northern Ireland*
H. Kleinpoppen, *Atomic Physics Laboratory, University of Stirling, Scotland*

Editorial Advisory Board

R. B. Bernstein *(New York, U.S.A.)*
J. C. Cohen-Tannoudji *(Paris, France)*
R. W. Crompton *(Canberra, Australia)*
J. N. Dodd *(Dunedin, New Zealand)*
G. F. Drukarev *(Leningrad, U.S.S.R.)*
W. Hanle *(Giessen, Germany)*
C. J. Joachain *(Brussels, Belgium)*
W. E. Lamb, Jr. *(Tucson, U.S.A.)*
P.-O. Löwdin *(Gainesville, U.S.A.)*
H. O. Lutz *(Bielefeld, Germany)*
M. R. C. McDowell *(London, U.K.)*
K. Takayanagi *(Tokyo, Japan)*

ATOM–MOLECULE COLLISION THEORY: A Guide for the Experimentalist
Edited by Richard B. Bernstein

ATOMIC INNER-SHELL PHYSICS
Edited by Bernd Crasemann

ATOMS IN ASTROPHYSICS
Edited by P. G. Burke, W. B. Eissner, D. G. Hummer, and I. C. Percival

AUTOIONIZATION: Recent Developments and Applications
Edited by Aaron Temkin

COHERENCE AND CORRELATION IN ATOMIC COLLISIONS
Edited by H. Kleinpoppen and J. F. Williams

COLLISIONS OF ELECTRONS WITH ATOMS AND MOLECULES
G. F. Drukarev

DENSITY MATRIX THEORY AND APPLICATIONS
Karl Blum

ELECTRON AND PHOTON INTERACTIONS WITH ATOMS
Edited by H. Kleinpoppen and M. R. C. McDowell

ELECTRON–ATOM AND ELECTRON–MOLECULE COLLISIONS
Edited by Juergen Hinze

ELECTRON-MOLECULE COLLISIONS
Edited by Isao Shimamura and Kazuo Takayanagi

INNER-SHELL AND X-RAY PHYSICS OF ATOMS AND SOLIDS
Edited by Derek J. Fabian, Hans Kleinpoppen, and Lewis M. Watson

INTRODUCTION TO THE THEORY OF LASER – ATOM INTERACTIONS
Marvin H. Mittleman

ISOTOPE SHIFTS IN ATOMIC SPECTRA
W. H. King

PROGRESS IN ATOMIC SPECTROSCOPY, Parts A, B, C, and D
Edited by W. Hanle, H. Kleinpoppen, and H. J. Beyer

THEORY OF MULTIPHOTON PROCESSES
Farhad H. M. Faisal

VARIATIONAL METHODS IN ELECTRON – ATOM SCATTERING THEORY
R. K. Nesbet

A Continuation Order Plan is available for this series. A continuation order will bring delivery of each new volume immediately upon publication. Volumes are billed only upon actual shipment. For further information please contact the publisher.

COLLISIONS OF ELECTRONS WITH ATOMS AND MOLECULES

G. F. Drukarev
Leningrad University
Leningrad, USSR

PLENUM PRESS • NEW YORK AND LONDON

Library of Congress Cataloging in Publication Data

Drukarev, Grigorii Filippovich.
Collisions of electrons with atoms and molecules.

(Physics of atoms and molecules)
Translation of: Stolknoveniia elektronov s atomami i molekulami.
Bibliography: p.
Includes index.
1. Collisions (Nuclear physics) 2. Scattering (Physics) 3. Electrons—Scattering. 4. Atoms.
I. Burke, P. G. II. Title. III. Series.
QC794.6.C6D7813 1987 539.7′54 85-16990
ISBN-13: 978-1-4612-8997-5 e-ISBN-13: 978-1-4613-1779-1
DOI: 10.1007/ 978-1-4613-1779-1

Softcover reprint of the hardcover 1st edition 1987
A Division of Plenum Publishing Corporation
233 Spring Street, New York, N.Y. 10013

Foreword

This book is a short outline of the present state of the theory of electron collisions with atomic particles - atoms, molecules and ions.

It is addressed to those who by nature of their work need detailed information about the cross sections of various processes of electron collisions with atomic particles: experimentalists working in plasma physics, optics, quantum electronics, atmospheric and space physics, etc.

Some of the cross sections have been measured. But in many important cases the only source of information is theoretical calculation. The numerous theoretical papers dealing with electronic collision processes contain various approximations. The interrelation between them and the level of their accuracy is often difficult to understand without a systematic study of the theory of atomic collisions, not to mention that theoretical considerations are necessary for the consistent interpretation of experimental results.

The main constituents of the book are:

1. General theory with special emphasis on the topics most important for understanding and discussing electron collisions with atomic particles.
2. Approximate methods for electron-atom collision cross section calculations. They are classified into 3 groups. The first of these are methods which allow the cross section of a given transition to be calculated with high accuracy in the energy region below the ionization threshold. These methods are capable of reproducing numerous resonance peaks and dips of width 10^{-2} - 10^{-3} eV in the energy

dependence of cross sections. The second are methods which allow smoothed-over resonance cross sections to be calculated with an accuracy up to a factor 1.5-2. The third are methods for estimation, within an order of magnitude, of the cross section at a given energy transfer independent of what happens with the atom.

3. Results of cross-section calculations for some particular cases for the following processes: elastic scattering of electrons by atoms and positive ions; excitation of atoms and positive ions; ionization; rotational and vibrational-rotational excitation of molecules; dissociative attachment; dissociative recombination; and dissociation through electronic excitation.

The third part of the book serves as an example of applications of the general theory and approximate methods. In addition to the results of particular calculations being interesting in themselves, they also reveal many important general features of collision processes. This aspect is emphasized in discussion of particular calculations.

Contents

SCATTERING OF A PARTICLE WITH SPIN: POLARIZATION PHENOMENA

THE SIMPLEST TWO-CHANNEL SYSTEM

COLLISIONS OF ELECTRONS WITH ATOMS AND IONS: GENERAL THEORY

APPROXIMATE METHODS FOR ELECTRON-ATOM COLLISIONS: CROSS SECTION CALCULATIONS

ELASTIC AND INELASTIC SCATTERING OF ELECTRONS BY ATOMS AND POSITIVE IONS

CONTENTS

Introduction

General Description of Electron Collision Processes with Atomic Particles

0.1 THE CONCEPT OF CROSS SECTION

Let us consider the collision of an electron with some target-atomic particle. Suppose that before the collision the electron as well as the target have certain momenta and projection of their spin on some specified fixed direction, and the target is in a certain state of internal motion specified by a set of quantum numbers. After the collision all momenta, spin projections and other quantum numbers will be changed (in general). Moreover the structure of the target can be changed (e.g., ionization).

The change of momenta and other quantities is governed by the conservation rules including conservation of the total energy and momenta of the colliding particles. The simplest form of these rules is achieved in the special coordinate system in which the center of mass is at rest. In the case of the collision of an electron with an atomic particle, the center of mass almost coincides with the center of the atomic particle because of the very small electron-to-proton mass ratio.

The relative probability of a certain change of momenta, spin, target quantum numbers, etc. is characterized by the differential cross section.

To define the cross section in terms of the measured quantities let us consider the typical arrangement of a collision experiment (Figure 0.1) (which is idealized in some respects). In the source S a beam of electrons is formed having energy E_0 and moving in the direction $\mathbf{n}_0$ with the spin projection μ on a certain fixed direction. The current density of the beam is I_0 electrons per sec per cm^2.

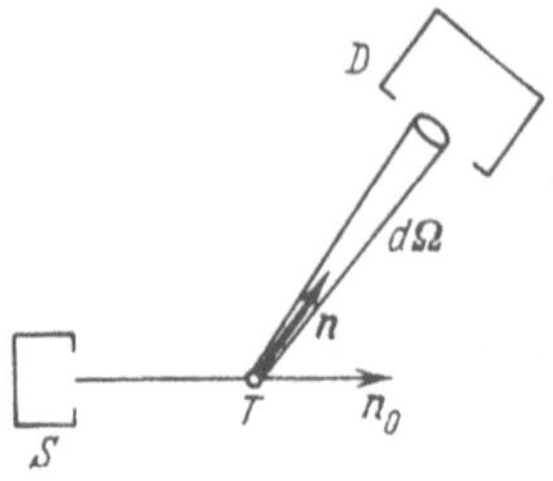

Fig. 0.1 Experimental set-up for cross section measurement.

The beam hits the target T containing N identical atomic particles in the same specified initial state. The detector D measures the part of the scattered electronic current dI going into the solid angle dΩ around the direction **n**. Let us assume that the energy of the scattered electrons and their spin projections can be measured. Then we can determine the current dI_{fi} which corresponds to a certain transition of atomic particles $i \to f$ (i stands for the initial set of quantum numbers and f for the final ones) induced by the electrons. If the density of the particles in the electron beam and in the target is low enough then dI_{fi} is proportional to I_0, $\mathbf{n}_0$ and dΩ:

$$dI_{fi} = Q_{fi}(\mathbf{n},\mathbf{n}_0,E_0)NI_0 d\Omega. \tag{1.1}$$

The ratio

$$dI_{fi}/I_0N = d\sigma_{fi}$$

is independent of the current density I_0 and the particle number in the target N. It has the dimension of the square of the length and is called the cross section for scattering in the solid angle dΩ for the transition $i \to f$. The cross section per unit solid angle $dI_{fi}/I_0Nd\Omega$ is called the differential cross section, $d\sigma_{fi}/d\Omega$.

From (1.1) we have

$$\frac{d\sigma_{fi}}{d\Omega} = Q_{fi}. \tag{1.2}$$

The total cross section is equal to the integral

$$\sigma_{fi} = \int Q_{fi} d\Omega \tag{1.3}$$

(the word "total" is often dropped).

According to general principles of quantum mechanics the differential cross section can be expressed as a square of the absolute value of a certain amplitude.

It is customary to introduce a factor which takes into account the obvious proportionality between the current density and the velocity of the particle v and write

$$Q_{fi} = \frac{v_f}{v_i} |A_{fi}(n,n_o)|^2. \tag{1.4}$$

The amplitude A_{fi} is the primary subject of a quantum mechanical calculation. If we want to indicate in explicit form all quantum numbers on which the amplitude depends, the Dirac bracket notation will be used:

$$A_{fi} = \langle a_f, b_f \ldots n|A|a_i, b_i \ldots n_o\rangle. \tag{1.5}$$

Here a_i, b_i ... are the quantum numbers of the initial states of the electron and the target (including spin projection) and a_f, b_f ... are those of the final.

The real experiment differs from the idealized set-up considered above in many respects. First of all, it is impossible to prepare the spins of all the electrons in a fixed state (the so-called "pure state" in a quantum mechanical sense). Only some degree of spin polarization can be achieved. The polarization **P** of an electron beam is an axial vector, its components P_x, P_y, P_z being the averaged - over the beam - value of the spin projections, $\langle\hat{s}_x\rangle$, $\langle\hat{s}_y\rangle$, $\langle\hat{s}_z\rangle$, divided by the absolute value $\hbar/2$ (in order to make the maximum polarization equal to unity):

$$P_i = \frac{2}{\hbar}\langle\hat{s}_i\rangle.$$

The beam is called unpolarized if **P** = 0. Methods of producing and analyzing the electron beam polarization are described in the book by Kessler (1976).

Next, sometimes the initial and final state of the atomic particle cannot be determined completely, etc. If there is a lack of selection in respect to the quantum number a, then the scattered current corresponds to the sum of currents over all final values a_f, averaged over the initial distribution of a_i.

The measured cross section will be

$$\bar{Q}_{fi} = \overline{\sum a_f Q_{fi}}$$

where the bar indicates the average over the initial distribution of a_i. In the case of a uniform distribution when all values of a_i have equal probability, the measured cross section $\bar{Q}_{fi}$ will be given by the expression

$$\bar{Q}_{fi} = \frac{1}{N_i} \sum_{a_i} \sum_{a_f} |\langle a_f b_f|A|a_i b_i\rangle|^2. \tag{1.6}$$

N_i is the number of possible values of a_i (provided it is finite). Moreover, there is always some spread over the relative velocity of the colliding particles. As a result the measured cross section σ is an average over this spread:

$$\sigma = \int \sigma_{fi} f(\mathbf{v}) d\mathbf{v} \tag{1.7}$$

Here $f(\mathbf{v})$ is the velocity distribution. Usually the function f is centered around some velocity $\mathbf{u}$, $f \equiv f(\mathbf{u} - \mathbf{v})$. So the averaged cross section will be a function of $\mathbf{u}$.

An alternative method of experimental study of electronic collisions is the observation of the radiation emitted by the excited atomic particle. It should be noted however that the observed radiation comes not only from one particular atomic transition, but sometimes from a series of cascade transitions.

0.2 INTERACTION OF ELECTRONS WITH ATOMS OR ATOMIC IONS

Let us consider the collision processes starting from very low energy electrons and passing to higher energy. Then we will pass from the simplest process to the more and more complicated. We begin with such a low energy that no excitation is possible. The only process which is possible is elastic scattering, leaving the target particle in its initial state.

If the electronic shells of the atomic particle are assumed to be frozen, then the problem reduces to electron scattering by a fixed field - so-called potential scattering. Although the real collision of an electron with the atomic particle is much more complicated than potential scattering, nevertheless the theory of potential scattering is very important in many respects and it will be considered in Chapter 1. Some complications are introduced into potential scattering by the existence of the electron magnetic moment and, as a consequence of this, the spin-orbit interaction. This interaction is a small relativistic effect, but it leads to interesting spin polarization phenomena. (The polarization phenomena are considered in Chapter 2.)

Let us consider briefly the factors which make the process of electron-atom elastic scattering more complicated than simple potential scattering. First of all, there is the perturbation of the atom by the scattered electron. This perturbation changes the atomic field which in turn influences the motion of the scattered electron and the scattering cross section.

When the scattered electron velocity is much less than characteristic velocity of the bound electrons in the atom v_0, then in the first approximation the above mentioned perturbation can be considered as the electric polarization of the atom by a point charge at rest. The change in the atomic field is represented by the interaction of a point charge with the induced electric dipole moment of the atom. Far away from the atom when the distance r is much larger than the Bohr radius a_0, the interaction energy is equal to

$$V = -\alpha/2r^4 \tag{2.1}$$

where α is the atomic polarizability.

We shall see later (Chapter 1, Section 1.5) that this interaction leads under certain conditions to a deep minimum in the energy dependence of the cross section, the so-called Ramsauer effect (Figure 0.2).

If we increase the electron velocity then the picture of polarization by a fixed charge at rest becomes inadequate. When the electron velocity approaches the characteristic atomic velocity v_0, then the interaction between the scattered electron and the atomic electrons leads to a new phenomena: formation of a transient negative atomic ion, which can be considered as a bound electron in the field of an excited atom. This is a quasistationary state because there is a possibility of its decay: the atom returns to the ground state and the electron is liberated.

The existence of such a state leads to the sharp resonance structure in the energy dependence of the scattering cross section (Figure 0.3). The position of the resonance on the energy scale E_r is equal to the difference between the excitation energy E_e and the bound energy of the captured electron W:

$$E_r = E_e - W. \tag{2.2}$$

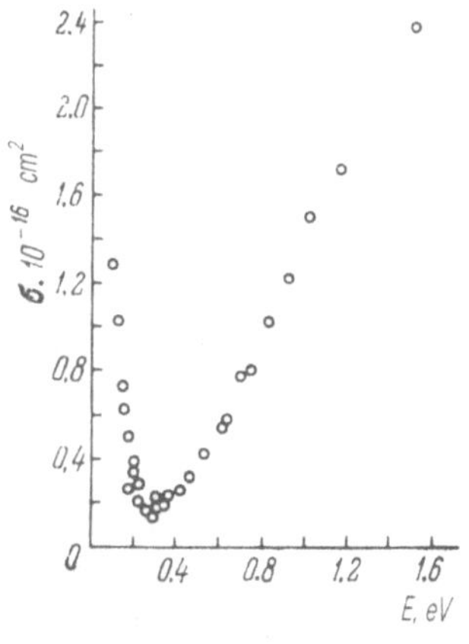

Fig. 0.2 The Ramsauer effect in Ar (Golden and Bandel, 1966).

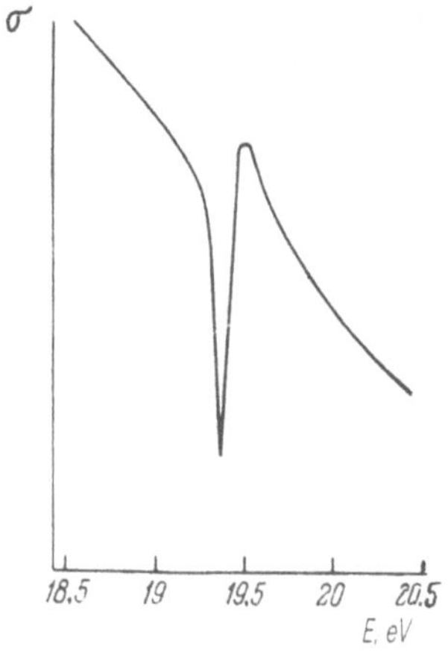

Fig. 0.3 Resonance in e - He scattering. The cross section is in arbitrary units (Schulz, 1973).

The width of the resonance Γ is related to the lifetime of the negative ion under consideration by the well-known equality:

$$\tau = \hbar/\Gamma. \tag{2.3}$$

A typical value of Γ is 10^{-3} - 10^{-2} eV. It is rather hard to observe such a resonance. One should spend much effort to reduce the spread of the energy in the electronic beam to such a small value as 10^{-2} - 10^{-3} eV. But if the spread is much larger than in the averaged spread-cross section there will be no trace of a resonance at all.

Another factor which complicates the scattering of an electron by an atom is the specific "exchange interaction" due to symmetry properties of the wave function of the whole system "electron + atom" with respect to exchange of its arguments. This interaction depends on the total spin of the system. Consequently the scattering cross section will also be spin-dependent. The simplest example is the scattering of an electron by a hydrogen atom. In this case we have to study the two-electron system which can be in two states: singlet (the total spin = 0) and triplet (the total spin = 1 in units of ħ). There will be two different amplitudes: singlet A^0 and triplet A^1. The difference between A^1 and A^0 leads also to spin polarization phenomena (Chapter 2).

To extract maximal information from the experiment one should ensure the selection of spin states both of the electronic beam (before and after the scattering) and of the target particle. In the absence of such selection the measured quantities will be in fact an average spin like (1.6). In the case of electron-hydrogen scattering the spin-averaged cross section is equal to

$$\sigma = \frac{3}{4}\sigma^1 + \frac{1}{4}\sigma^0. \tag{2.4}$$

Continuing to raise the electron energy we cross the excitation threshold, after which two processes are possible: elastic scattering and excitation. Let us briefly review some general features of the excitation process.

Unlike the transitions induced by light there are not any selection rules. All transitions are allowed including those which change the atomic spin, provided the energy is sufficient.

Like the case of elastic scattering the cross section depends on total spin (through the symmetry properties of the wave function).

At threshold the amplitude A_{fi} is finite for neutral atoms. Consequently, the excitation cross section at threshold will be zero. In the vicinity of the threshold where A_{fi} is practically constant the cross section will be proportional to v_f.

The only exception is the hydrogen atom. Due to degeneracy of the energy levels with respect to the orbital quantum number ℓ, there is a permanent dipole moment in each excited state. As a result, the excitation cross section will be finite at the threshold.

In the positive atomic ion case, because of the Coulomb field the excitation cross section will be finite at the threshold.

The excitation of an atom is the result of a change in the whole system "electron + atom." The transition in the atom and the change in the scattered electron motion are mutually dependent on each other. Like the elastic scattering, this mutual influence can be pictured as a polarization of the excited atom by an outgoing electron when it is slow. Near the next excitation threshold again a transient negative ion can be formed which leads to resonances in excitation cross section.

All this makes the energy dependence of excitation cross section look rather complicated.

As an example the excitation cross section of the 2s-state of the hydrogen atom is shown in Figure 0.4 in the energy region below the n = 3 threshold. It is seen that by averaging the sharp resonances are smeared. Just this smoothed cross section is observed actually. Finally, one should mention the important factor - conservation of the total number of electrons during the excitation of the atom. Because of this, the cross sections of various processes (elastic scattering and excitation of all excited states

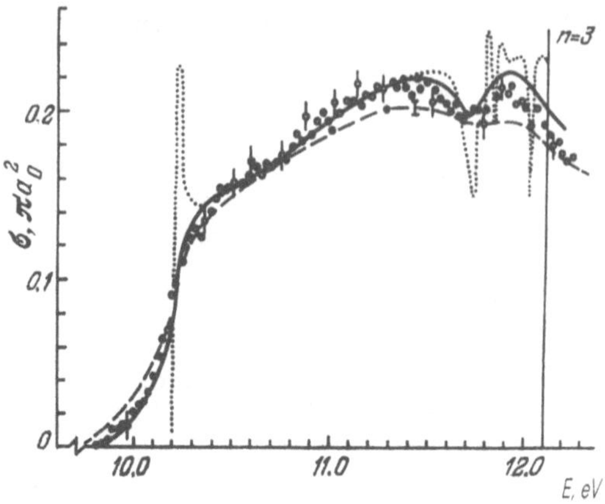

Fig. 0.4 2s-state excitation cross section in H. Dotted line - theory. Dashed line - theoretical result averaged over energy spread 0.1 eV. Full line and circles - measurements (Burke et al., 1967).

which are allowed at a given energy) are interrelated. Moreover, this interrelation leads to interesting features in the elastic scattering differential cross section just at the threshold - so-called cusps (Figure 0.5).

Raising the electronic energy still more we cross the ionization threshold, after which ionization becomes possible.

The ionization is described by the differential cross section:

$$d^3\sigma = \frac{v_f}{v_i} |A_{fi}|^2 v'^2 dv' d\Omega d\Omega'. \tag{2.5}$$

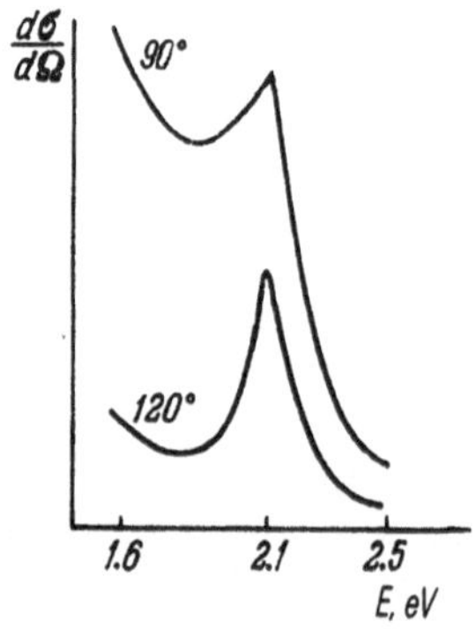

Fig. 0.5 A cusp in e - Na scattering (Andrick et al., 1972).

Here $d\Omega$ is the solid angle in which the incident electron moves after the ionization, $d\Omega'$ is the solid angle in which the ejected electron moves, the velocity of this electron being between v' and $v' + dv'$. Now, we should take into account that electrons are indistinguishable and the motion of the incident electron in $d\Omega$ and the ejected electron in $d\Omega'$ is identical with the motion of the incident electron in $d\Omega'$ and the ejected in $d\Omega$. So we should not talk about the incident and ejected electrons after the ionization occurs. The only real difference between these two electrons is their velocity. One is slow, another is fast. Let us put by definition that $v_f \geqslant v'$. Then we will say that the cross section (2.5) describes the process in which the fast electron moves in the solid angle $d\Omega$ with the velocity v_f, and the slow one in the solid angle $d\Omega'$ with the velocity between v' and $v' + dv'$. The total cross section is given by the integral

$$\sigma = \frac{v_f}{v_i} \int d\Omega \int d\Omega' \int_0^{v'_{max}} v'^2 dv' |A_{fi}|^2. \tag{2.6}$$

Among the various interesting details of the ionization process the threshold behavior should be mentioned. The total ionization cross section varies with the electronic energy E, as in $(E - I)^\alpha$, where I is the binding energy of the atomic electron, and $\alpha \approx 1.127$ (see Section 7.1). This threshold law was verified in several experiments. An example is shown in Figure 0.6. Here the measured derivative of the ionization cross section $d\sigma/dE$ is plotted as a function of the incident electron energy E. One can easily recognize the peaks indicating the excitation of discrete energy states of the atom. The smooth curve for the continuum spectrum corresponds to $\alpha = 1.127$.

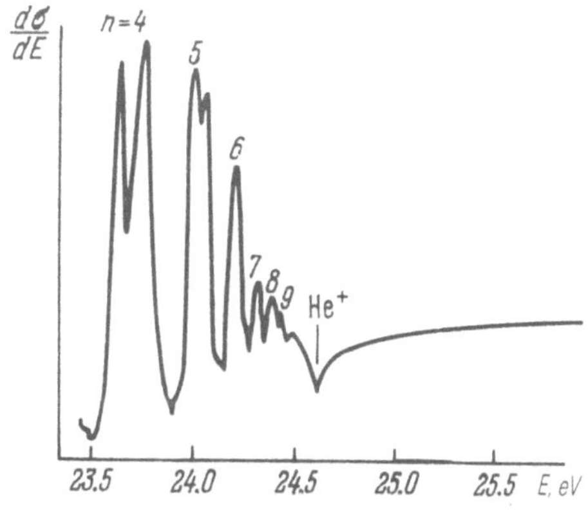

Fig. 0.6 $d\sigma/dE$ (arbitrary units) for the e - He inelastic cross section near the ionization threshold (Cvejanović and Read, 1974).

Similar to ionization is electron detachment from a negative atomic ion. The essential difference is that both detached and scattered electrons move in the field of a neutral atom, while in the case of atom ionization they move in a Coulomb field of the ion. This leads to a significant difference in threshold behavior.

0.3 SOME INFORMATION ABOUT ATOMIC STRUCTURE

To make the discussion of various processes more explicit, we give here a short summary of some facts and concepts about atomic structure. (More detailed and systematic accounts can be found in textbooks on quantum mechanics and special monographs on atomic structure.)

In the theoretical treatment of the structure and properties of atoms one can recognize different levels of accuracy, depending on how complete and in what approximation the Coulomb interaction between electrons and the relativistic effects are taken into account.

Let us begin with the simplest case - the one electron hydrogen atom or hydrogen-like ion. In the most crude approximation the theory of a hydrogen-ion system reduces to the problem of a charged particle in the Coulomb field of the nucleus $-Ze/r$ - neglecting all relativistic effects. The energy levels (Figure 0.7a) are given by the well-known expression

$$E_n = -\frac{mZ^2e^4}{2\hbar^2n^2} . \qquad (3.1)$$

From (3.1) it follows that for large n the distance between adjacent levels decreases by n^{-3} and, say, for $n = 10$ is equal to 10^{-2} eV. The electronic wave function is an eigenfunction of the operators of the energy H, angular momentum $\hat{L}$ and its projection $\hat{L}_z$ on some direction taken as a z-axis:

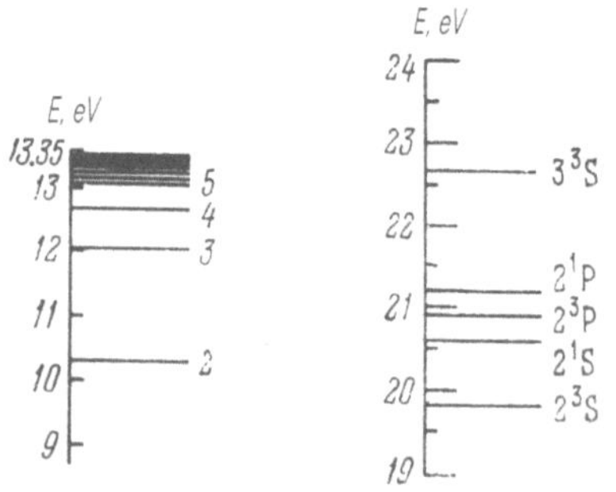

(a) (b)

Fig. 0.7 A part of the energy spectrum in H and He (standard spectroscopic symbols S,P... instead of $\ell = 0,1$ ... are used).

$$H\psi_{n\ell m} = E_n\psi_{n\ell m}$$

$$\hat{L}^2\psi_{n\ell m} = \hbar^2\ell(\ell+1)\psi_{n\ell m} \qquad (3.2)$$

$$\hat{L}_z\psi_{n\ell m} = \hbar m\psi_{n\ell m}.$$

For a given ℓ the quantum number m can have all integer values between $-\ell$ and ℓ. The quantum number ℓ for given n can have all integer values in the interval 0... n − 1. There are analytical expressions for the wave functions $\psi_{n\ell m}$ which we will not reproduce here. The wave function depends on 3 quantum numbers: n, ℓ, m; while the energy depends only on n. Such a situation is called degeneracy. The degeneracy with respect to the quantum number m is an obvious result for any spherical-symmetric potential. But the degeneracy with respect to the quantum number ℓ is by no means "obvious" and it is a special property of the Coulomb field. Just this degeneracy is responsible for the existence of an electric dipole moment of an atom in excited state.

So far we have not considered the spin of the electron. Now let us discuss how to include spin quantum numbers in the wave function.

Denote by $\hat{s}$ the spin operator and by $\hat{J}$ the total angular momentum operator, which is the sum

$$\hat{J} = \hat{L} + \hat{s}.$$

The eigenvalues of $\hat{s}^2$ and $\hat{s}_z$ are $\hbar^2 s(s+1)$ and $\hbar\mu$ respectively. In our case $s = \frac{1}{2}$, $\mu = \pm\frac{1}{2}$. The eigenvalues of $\hat{J}^2$ are $\hbar^2 j(j+1)$ where $j = |\ell - \frac{1}{2}|$, $\ell + \frac{1}{2}$ for given ℓ, and the eigenvalues of J_z are $\hbar m_j$ where $m_j = m \pm \frac{1}{2}$. There are two alternative noncommuting ways of numbering the states. In the first, the state is identified by the quantum numbers ℓ, m, μ. In the second, the state is identified by the quantum numbers j, m_j, ℓ (we do not indicate the value of $\hat{s}^2$ because in our case it is fixed). Denote by u_μ the eigenfunction of $\hat{s}_z$ corresponding to eigenvalue $\hbar\mu$. Then the product $\Phi_{n\ell m\mu} = u_\mu\psi_{n\ell m}$ will be the total wave function in the ℓ, m, μ-representation. The total wave function in j, m_j, ℓ will be:

$$\Psi_{nj\ell m_j} = \sum_{\substack{m,\mu \\ (m+\mu)=m_j}} \Phi_{n\ell m\mu} C^{jm_j}_{\ell m;\frac{1}{2}\mu} \qquad (3.4)$$

where $C^{jm_j}_{\ell m;\frac{1}{2}\mu}$ are the Clebsch-Gordan coefficients.

The more accurate description of the atom will be achieved if we take into account the relativistic effects. Up to the order $\sim (v/c)^2$ the numbers n, j, ℓ, mj will still be "good quantum numbers" and in the first approximation the wave function (3.4) can be used.

In this approximation the relativistic correction to the energy level (3.1) is given by the expression

$$\Delta E_n = E_n \frac{\alpha^2 Z^2}{n} \left(\frac{1}{j+\frac{1}{2}} - \frac{3}{4n}\right) \tag{3.5}$$

where $\alpha = e^2/\hbar c \approx 1/137$ is the fine-structure constant. Now the energy depends not only on n, but also on j which can have the values $\frac{1}{2}, \ldots, n-\frac{1}{2}$. So instead of one level in the nonrelativistic limit we get a set of n levels (Figure 0.8). This splitting is known as fine structure. For n = 1 there is no splitting but rather a shift of the single level. There is still some degeneracy because levels with $\ell = j + \frac{1}{2}$ and $\ell = j - \frac{1}{2}$ have the same energy. Notice that $E_n \sim Z^2$ while $\Delta E_n \sim Z^4$. Thus, the ratio

$$\Delta E_n/E_n \sim Z^2.$$

The next stage of accuracy is achieved by taking into account specific quantum electrodynamical effects (vacuum polarization, etc.). Reaching this stage one is forced to introduce corrections to the wave function $\Psi_{nj\ell m_j}$, which involves the functions $\Psi_{\ell-1}$ and $\Psi_{\ell+1}$. The splitting and shift of the levels due to such effects is much less than the fine structure (the Lamb shift is an example). We will not consider such effects in this book.

Let us turn now to a many-electron atom. In the nonrelativistic limit its state is identified by the total orbital angular momentum quantum number L and total spin quantum number S. The energy levels corresponding to given L and S are called atomic terms. The dependence of energy on spin function comes from the dependence of the symmetry properties of the coordinate part of the wave function on the total spin and from the fact that the solutions of the Schrödinger equation with different symmetry properties correspond to different energies. The simplest example is the He atom. There are singlet (S = 0) and triplet (S = 1) states. Several terms are shown on Figure 0.7b, together with their spectroscopic symbols.

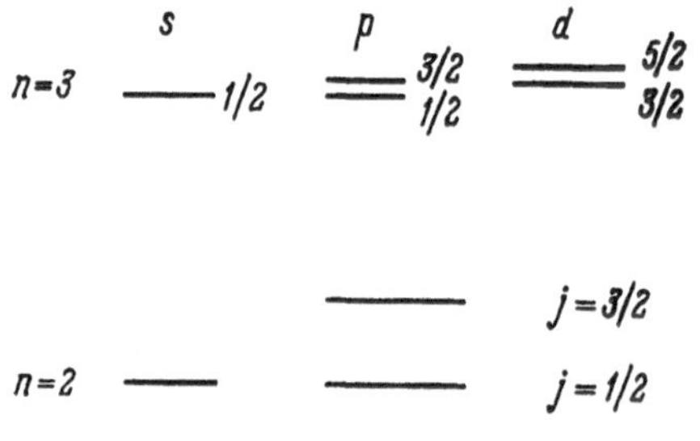

Fig. 0.8 Fine structure splitting of n = 2 and n = 3 energy levels in H.

The total wave function Φ of the atom (which depends both on coordinates of electrons and projection of their spin) is an eigenfunction of the operators H, $\hat{L}^2$, $\hat{S}^2$, and $\hat{L}_z$, $\hat{S}_z$

$$\begin{aligned} H\Phi &= E\Phi \\ \hat{L}^2\Phi &= \hbar^2 L(L+1)\Phi \\ \hat{L}_z\Phi &= \hbar\mu\Phi \\ \hat{S}^2\Phi &= \hbar^2 S(S+1)\Phi \\ \hat{S}_z\Phi &= \hbar\mu_s\Phi. \end{aligned} \tag{3.6}$$

The function Φ should be also antisymmetric with respect to the exchange of coordinates and spin quantum number of any pair of electrons:

$$\Phi(\ldots\ r_p\mu_p\ \ldots\ r_q\mu_q\ \ldots) = -\Phi(\ldots\ r_q\mu_q\ \ldots\ r_p\mu_p\ \ldots). \tag{3.7}$$

The consequence of this antisymmetry is the Pauli principle. (More details of symmetry properties and construction of the wave function corresponding to certain values of S are given in Chapter 2).

Unlike the hydrogen atom there are no exact analytical expressions for Φ. It should be realized that any attempt at straightforward numerical solution of the Schrödinger equation for a many-electron atom by means of a computer would be useless. The wave function Φ for a certain energy level of an N-electron atom depends on 3N coordinates. If we try to prepare a table of the numerical values of Φ selecting for each coordinate only p points, then the table will contain p^{3N} numbers. Let us consider an example: $p = 10$, $N = 5$. There will be 10^{15} numbers in the table. That is to say, $\sim 10^9$ volumes with about several hundred pages each! And all this for only one given energy level! Even if such a table could be produced it would be impossible to use it.

A good starting point to overcome the difficulty is the independent particle model. In the most simple version of this model the Coulomb interaction between electrons is neglected completely. Then each electron moves independently from any other in the Coulomb field of the nucleus having the energy levels according to (3.1). The wave function is built up from the one-electron wave functions of the familiar hydrogen atom problem. However, this approximation is too crude. A more refined version is achieved by the self-consistent field method. Here each electron moves independently from the others in the Coulomb field of the nucleus and the averaged field of other electrons (a part of the electron interaction is left still neglected in this approach). For any electron the quantum numbers ℓ, m_ℓ and the energy can be indicated. To distinguish states with the same ℓ but different energy, the quantum number n is used. It is agreed that $n \geqslant \ell$ in analogy to a hydrogen atom. Because the averaged field of electrons is not the Coulomb field of a point charge, there will be no degeneracy with respect to quantum number ℓ. As a result the energy of an electron will depend both on n and ℓ. It can happen that the energy of an electron with quantum numbers n_1, ℓ_1 will be higher than that with n_2, ℓ_2, in spite of n_1 being smaller than n_2.

The pair of quantum numbers n, ℓ is called a configuration. For given ℓ there are $2\ell + 1$ possible values of m_ℓ. Taking into account two possible values of the spin projection we have $2(2\ell + 1)$ different states for the same n and ℓ. They are called equivalent. According to the Pauli principle in each of the equivalent states there can be no more than one electron. The maximal number of electrons with the same n and ℓ is just $2(2\ell + 1)$. They form an electronic shell. Then if N electrons have the same configuration it is agreed to write $(n\ell)^N$. The symbol of a shell is then $(n\ell)^{2(2\ell+1)}$. For n = 1,2,3 ... we have the shells $(1s)^2$, $(2s)^2$, $(2p)^6$, $(3s)^2$, $(3p)^6$, $(3d)^{10}$.... Fixing a certain configuration and using the rules of addition of the angular momentum, and also taking into account the Pauli principle, one can find the possible values of quantum numbers L and S for all atomic terms corresponding to one and the same configuration. The wave function Φ in this method is expressed as a linear combination of one-electron coordinate functions $\phi_{n\ell}(r)$ and spin function u_s, subjected to the antisymmetry condition. The simplest form of such an antisymmetric function is the determinant built up from the functions $\phi_{n\ell}(r)u_s$ - the so-called Slater determinant. Now, one should not forget that the wave function Φ must be an eigenfunction of $\hat{S}^2$. The expression of the wave function in the form of a single Slater determinant is not always compatible with this requirement. If this is the case, a linear combination of determinants should be used. An excited state of the He atom is an example. To find the one-electron function $\phi_{n\ell}$, Fock (1930) used the variational principle of quantum mechanics according to which the expression

$$\int \Phi^* H \Phi dr_1 \ldots dr_n$$

should be stationary with respect to a small variation of the functions $\phi_{n\ell}$ (of which Φ is composed). In fact this stationary state should be a minimum. From this condition, Fock derived a set of equations which are the basis of the self-consistent field method. If instead of the determinant form of Φ one uses a simple product of one electronic function $\phi_{n\ell}$, then the equations derived for $\phi_{n\ell}$ will become much more simple and in fact coincide with the equations proposed by Hartree (1928). For this reason the self-consistent field method is often called the Hartree-Fock method.

To solve the Hartree-Fock equations, very extensive computational work is needed. Nevertheless, for many atoms this work was done recently by several authors and the results in tabulation form can be found in the literature. In order to reduce computational work and make the functions $\phi_{n\ell}$ more convenient for further use, approximate analytical expressions of the form

$$\sum_n c_n r^{\beta_n} \exp(-\alpha_n r)$$

are used, the coefficients c_n, β_n, α_n being determined by means of the variational principle (including the necessary orthogonality conditions). The table of numerical values of c_n, β_n, α_n for many states also can be found in the literature.

The excitation of an atom according to the independent particle model is a transition of just one electron from one state to another, unoccupied state. Consider the excited states which come from the transition of one of the electrons belonging to the outer shell. The lowest excited states are very different in different atoms and there is very little in common for them. But all the high-excited states in all atoms are very like the excited states of the hydrogen atom. The energy of an electron which is in a high-excited state can be represented in "hydrogen-like" form

$$E_n = -\frac{me^4}{2\hbar^2}\frac{1}{(n-\Delta_\ell)^2} \tag{3.8}$$

where Δ_ℓ is the so-called quantum defect. Another kind of excited state is due to transitions from the inner shell. In most cases such a state is above the ionization threshold of the atoms. Nevertheless, as far as the electrons are considered independently this state will be a stationary one.

The next stage of accuracy in the theoretical description of a many-electron atom (still in the nonrelativistic limit) is reached if we take into account that part of the interaction between electrons which is neglected in the previous treatment. That involves consideration of the electronic correlations. This effect can be treated in many ways. One of them is the method of superposition of configurations. Another approach is the use of modern many-body theory methods. The interelectronic correlation affects in some way the energy and the atomic wave function for the stable states. But the most important change comes for such excited states which are above the ionization threshold. Because of electronic correlation they become unstable, which leads to the ionization of the atom.

Finally we will mention the rather complicated problem of addition of electronic orbital and spin momenta and formation of total angular momentum of the atom. The simplest situation occurs when the relativistic effects are small. Then all spin momenta of the electrons combine together and form the total momentum S. All orbital momenta give the total orbital momentum L. In this case the atom is characterized by the quantum numbers J, L, S. Then similar to (3.4) we can write the connection between L, M, S, μ_s representation and $JLSM_J$ representation

$$\Psi_{LSJM_J} = \sum_{M\mu_s} C^{JM_J}_{LM;S\mu_s} \Phi_{LMS\mu_s}. \tag{3.9}$$

0.4 GENERAL CHARACTERISTICS OF THE CROSS SECTION CALCULATION METHODS

The methods for calculating cross sections will be considered in detail in Chapter 5. In this section we will outline the basic ideas behind the methods.

There are three kinds of methods. One of them attempts to calculate the cross sections including the effect of polarization of the atom, resonances due to the transient negative ions, and various threshold effects with a high accuracy.

The most popular among them is the close-coupling method. The total wave function of the system "electron + atom" is represented as a superposition of the ground state and all excited state wave functions of the atom. The coefficients in this superposition are functions of the scattered electron coordinates and describe the interaction of the scattered electron with the atom. For these functions a system of integro-differential equations can be obtained. Strictly speaking, the number of equations in this system is infinite, but to a certain approximation this system can be replaced by the finite one, neglecting the rest. Using modern computers, such a finite system of coupled equations can be solved. One can hope to get a satisfactory approximation - the better, the more equations remaining in the system. This method ensures that the interrelation between the cross sections of various processes will hold (at least approximately). It is a great advantage of this method. However, practical application of this method involves a very great amount of computational work. The number of equations in the system should be at least the same as the number of open channels. Because of this, only the excitation of the lowest excited states is practically tractable; it is a disadvantage of this method. To overcome it some modifications of the method were developed. It should be noticed that the atomic wave functions which enter into the equations must be sufficiently accurate; otherwise the solution of the close coupled equations will be no more than a waste of time.

The aim of another kind of method is the calculation of the smoothed-over resonance cross sections. The most widely used is the Born approximation and some of its modifications. In the Born approximation the interaction between the scattered electron and the atom is treated as a small perturbation. The transition amplitude is expressed as an integral which contains the atomic wave functions in the initial and final states. The evaluation of an integral is much easier than that of the solution of a system of coupled integro-differential equations. But the cross sections of various processes calculated in this approximation will not be in general interrelated properly. Another shortcoming of the Born approximation is that the scattering amplitude from two target particles will be (in this approximation) a sum of amplitudes from each particle separately. Actually there are multiple-scattering effects and the amplitudes will not be additive. This shortcoming is partly overcome in the Glauber (or eikonal) approximation.

The treatment of the interaction between the electron and an atom as a small perturbation seems quite reasonable at high energy. One can expect the Born approximation to be valid when the energy is high enough, but surprisingly good results were obtained even at low energy near the threshold, where one cannot consider the electron-atom interaction as a small perturbation. The calculated cross section differs from the observed averaged resonance by a factor 1.5 - 2 (but not by several orders of magnitude as one could expect). It so happens that the errors due to the perturbation treatment have the tendency to compensate each other. This lucky property of the Born approximation makes it a suitable basis for the construction of semiempirical and interpolation expressions. More consistent and accurate methods for calculation of smoothed cross sections are based on the simplification of the close-coupling method (so-called distorted waves approximation). The Born approximation gives rather good results if it is applied to the total cross section computation, but for the differential cross section the results can be sometimes quite erroneous, especially at large angles. A second-order perturbation correction in this case brings an improvement.

The last kind of method is intended to estimate the cross section of a given energy transfer rather than to estimate the cross section of a given transition. The estimation of the energy transfer can be quite successfully made by means of classical mechanics. Especially good results were obtained for ionization. Even the binary collision approximation which takes into account only the collision between the scattered and atomic electrons (neglecting the interaction with the nucleus) gives the ionization cross section within the error $\sim$ 50%.

The simplest approximations like Born or binary classical are a good starting point to deduce various scaling relations. Of course there are no scaling relations between the exact cross sections, but there is an approximate scaling. To give an example, let us consider ionization. According to the binary classical approximation the product of the cross section with the square of the ionization potential $\sigma \cdot I^2$ is a universal function of the ratio E/I, E being the electron energy. Now, if we calculate σI^2 using the experimental values for σ and I and plot it against E/I, then the results for various atoms will not lie on one and the same curve. However, they will be all rather near the universal curve and the spread will not be very large.

0.5 COLLISION PROCESSES OF ELECTRONS WITH MOLECULES AND MOLECULAR IONS

There is a great variety of collision processes due to various types of excitation. As an example let us consider the H_2 molecule. Owing to the small ratio of electronic mass to the mass of the nucleus, the theoretical treatment of molecular structure is split approximately into two parts.

First, the electronic energy E is determined in the field of the nuclei at rest. The energy E depends on the distance between the nuclei, R. Adding to the electronic energy E(R) the Coulomb repulsion energy between the nuclei e^2/R, we get the total energy U(R) of the molecule in this approximation. The functions U(R) are called the electronic terms of the molecule. In the limit $R \to \infty$ the value of U approaches the sum of the energy of isolated hydrogen atoms. At a finite distance the interaction between the atoms comes into play. Due to the dependence of the symmetry of the wave function on the total spin of the system there will be two terms - singlet and triplet - which originated from the same atomic states. In Figure 0.9 the terms which originated from the ground states of both atoms are shown. There are indeed other terms which originated from excited states of one or both atoms.

The next step is the consideration of nuclear motion. The electronic term U(R) plays the role of the potential energy of nuclear interaction. To determine the nuclear motion one should solve the Schrödinger equation with the potential energy U. From Figure 0.9 one can deduce that only the singlet term forms a potential

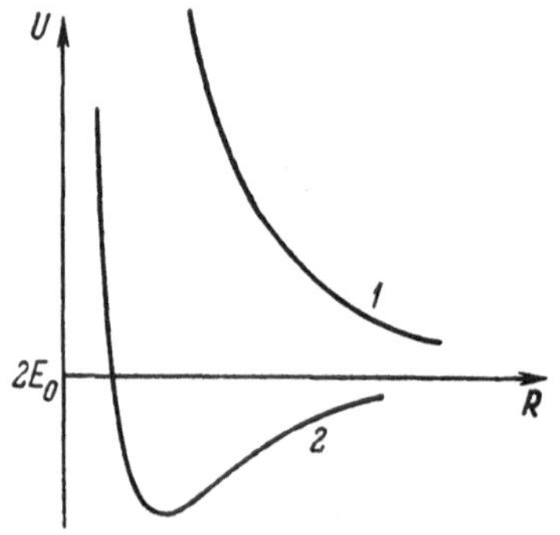

Fig. 0.9 Potential energy curves in H_2: 1 - triplet state and 2 - singlet state.

well in which a bound state is possible. The potential well in this case is spherically symmetric. Then the angular momentum of nuclear relative motion is an integral of motion and therefore one can characterize the stationary state by a corresponding quantum number j. There can be several states with the same j but different energy. To distinguish one from another, one more quantum number is introduced and it is denoted by v. The energy of low excited states of the nuclear motion can be expressed approximately in the form

$$E = E_0 + Bj(j+1) + \hbar\omega(v+\tfrac{1}{2}).$$

The term $\hbar\omega(v+\frac{1}{2})$ is like the harmonic oscillator energy and the term $Bj(j+1)$ is like the rigid rotator energy. Therefore it is said that the term $\hbar\omega(v+\frac{1}{2})$ corresponds to the vibrational levels, and the term $Bj(j+1)$ to the rotational energy levels.

If ε is the order of magnitude of the distance between the electronic excited states, then the distance between the vibrational levels is $\sqrt{m/M}\varepsilon$, and between the rotational levels, is $m/M\varepsilon$ where m is the electronic mass and M is the nuclear mass. However, for the high-excited states the above approximate expression for the nuclear motion energy becomes invalid. It follows that the words "rotational" and "vibrational" should not be considered too literally.

Let us consider now the possible processes of electron-molecule collisions. Below the excitation threshold of the rotation level excitation, the only possible process is elastic scattering. At such a low energy the velocity of an electron becomes comparable with that of the nuclei. As a result it is not possible in this case to separate the nuclear motion from the electronic, which makes the problem very difficult. The first inelastic process which becomes possible as we increase the electronic energy is the excitation of rotation. The rotational excitation is due to the interaction of the electron with the quadrupole or dipole momentum of the molecule. Increasing further the electronic energy we cross the vibrational excitation threshold, after which the vibrational excitation becomes possible. This kind of excitation proceeds through the formation of a transient negative molecular ion and subsequent decay in some excited state of the nuclear motion. Depending on the ratio of the lifetime of this transient ion and vibrational period, there are two possible shapes of the cross section versus energy curve (Figure 0.10 and Figure 0.11). On Figure 0.10 the vibrational excitation cross section of the H_2 molecule is plotted. The lifetime of the transient ion in this case is short. On Figure 0.11 the vibrational excitation of the N_2 molecule is plotted. This is an example of the long-lived transient ion.

There are also various dissociation processes. One of them is dissociative attachment, e.g.,

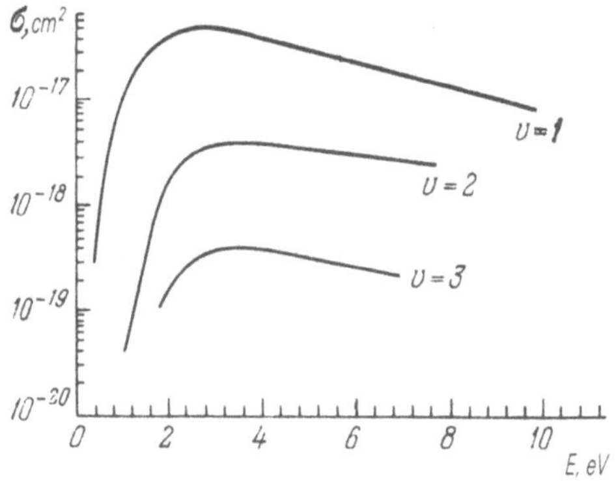

Fig. 0.10 Vibrational excitation cross section in H_2 (Schulz, 1973).

$$e + H_2 \rightarrow H + H^-$$

which is just another channel of decay of the transient negative molecular ion. Another process is dissociative recombination, e.g.,

$$e + H_2^+ \rightarrow H + H.$$

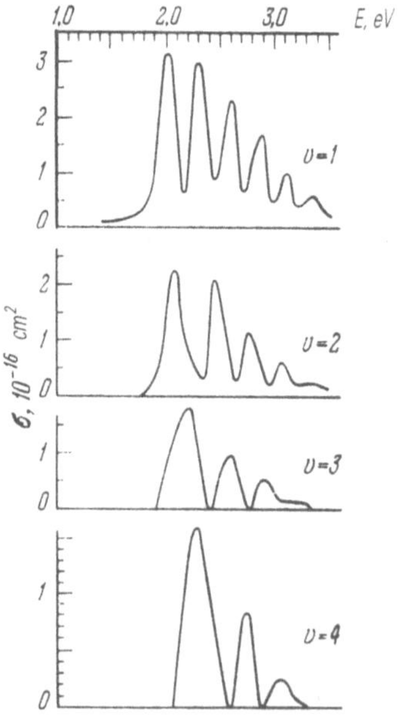

Fig. 0.11 Vibrational excitation cross section in N_2 (Schulz, 1973).

The dissociation also can proceed through the excitation of the repulsive electronic term, e.g., the triplet term for H_2 in Figure 0.9:

$$e + H_2 \rightarrow e + H + H.$$

Actually, the molecules as a rule are before the collision in excited rotational and sometimes even vibrational states. Then the average over the initial distribution cross sections is needed.

It follows that the theory of electron-molecule collisions is in many respects much more complicated than the case of the electron-atom collisions.

1
Scattering of a Particle by a Potential

1.1 THE WAVE FUNCTION. SCATTERING AMPLITUDE

The most consistent way of describing the scattering process would be to follow the time evolution of the particle wave function, starting with a certain initial form of it. This initial form corresponds to a particle localized at a given time in the source far from the target. The time evolution of the wave function is governed by the Schrödinger equation

$$i\hbar \frac{\partial \Psi}{\partial t} = H\Psi .$$

Because of the space and time localization of the particle in the source, there will be a quantum uncertainty in the energy and momenta. (We do not consider at this moment the macroscopic spread in energy and momentum due to the thermal motion in the source, etc.)

Usually the quantum uncertainty of the particle energy in any real scattering experiment is very small. Thus, it seems quite reasonable to make an idealization and assume that the energy has a sharp value E. We can then use the stationary state in which the wave function Ψ is factorized:

$$\Psi = \psi(r)e^{-i\frac{Et}{\hbar}} .$$

The function ψ satisfies the equation

$$\Delta\psi + (k^2 - V)\psi = 0 \tag{1.1}$$

which does not contain the time at all. Here $k^2 = 2mE/\hbar^2$; $V = 2mU/\hbar^2$. U is the potential energy of the particle. We will suppose that it depends on the distance from a certain center only and decreases when the distance is increasing. The position of the potential center will be taken as the origin of the coordinate system. We will use the atomic system of units in which the electronic mass, its charge and the quantum constant $\hbar$ are taken as unity.

It is obvious that now we are unable to follow in detail the time evolution from the inital state of the particle in the source to the final one.

Instead we should introduce some additional information about the function ψ. Such information, which replaces the more consistent time-dependent treatment of the scattering process, is provided by the asymptotic behavior of ψ at large r which serves as a boundary condition. We assume that far away from the potential center, at $r \to \infty$,

$$\psi \sim e^{ik\mathbf{n}_0 \cdot \mathbf{r}} + A(\mathbf{n},\mathbf{n}_0)e^{ikr}/r. \tag{1.2}$$

$$\mathbf{n}_0 = \mathbf{k}/k, \ \mathbf{n} = \mathbf{r}/r$$

The first term in (1.2) which is a plane wave describes the motion of an incident particle (in the absence of U) with the momentum **k**. The second term, which is an outgoing spherical wave, describes the motion of the scattered particle in this or some other direction with the relative probability determined by the amplitude A.

It should be noticed that in any real scattering experiment the incident beam is collimated and, strictly speaking, it is not an ideal plane wave. But the aperture is rather wide on the atomic scale so there is no need to indicate explicitly in (1.1) or (1.2) the action of the collimator.

It seems obvious that a shift of the origin of coordinates does not produce any real observable effect. However, it brings an additional phase factor in the amplitude $\exp[ik(\mathbf{n}_0 - \mathbf{n})\cdot \mathbf{a}]$ where **a** is the shift vector.

Let us establish now the relation between the amplitude A and the scattering cross section. We calculate the outgoing probability current density using the familiar expression

$$j = \frac{i}{2}(\psi \operatorname{grad}\psi^* - \psi^* \operatorname{grad}\psi) \tag{1.3}$$

and inserting the outgoing part of (1.2), $A/r \exp(ikr)$. For the radial component of the current density we have

$$j_r = |A|^2 \frac{k}{r^2} .$$

Therefore the current across the surface element dS will be

$$dj_r = |A|^2 k \frac{dS}{r^2} = |A|^2 k d\Omega. \tag{1.4}$$

Dividing this by the current density of the incident beam $j_0 = k$, we will obtain by definition the cross section

$$d\sigma = |A|^2 d\Omega. \tag{1.5}$$

In the general case of a potential which depends on all three cartesian coordinates the amplitude A as well as the cross section depends both on $\mathbf{n}$ and on $\mathbf{n}_0$. But in the case of a central field when the potential energy U depends only on distance r from the center, A depends on the angle between $\mathbf{n}_0$ and $\mathbf{n}$ only. Besides $\mathbf{n}$ and $\mathbf{n}_0$, the amplitude depends on the particle energy and also implicitly on the shape of U.

The aim of the theory is to establish how the characteristic features of the scattering amplitude A depends on all these factors.

An expression for the amplitude A in terms of the wave function ψ can be constructed which is convenient for general discussion. To derive it, let us transform the differential equation (1.1) into an integral equation taking into account (1.2).

$$\psi = e^{ik\mathbf{n}_0\cdot\mathbf{r}} - \frac{1}{4\pi}\int \frac{e^{ik|\mathbf{r}-\mathbf{r}'|}}{|\mathbf{r}-\mathbf{r}'|} V(\mathbf{r}')\psi(\mathbf{r}')d\mathbf{r}'.$$

It is easy to verify this equation using the relations

$$(\Delta + k^2)e^{ik\mathbf{n}_0\cdot\mathbf{r}} = 0, \quad (\Delta + k^2)\frac{e^{ik|\mathbf{r}-\mathbf{r}'|}}{|\mathbf{r}-\mathbf{r}'|} = -4\pi\delta(\mathbf{r}-\mathbf{r}').$$

At $r \to \infty$ one can neglect $\mathbf{r}'$ in comparison with $\mathbf{r}$ in the denominator of the integrand and in the exponent expand $|\mathbf{r} - \mathbf{r}'|$ in powers of $\mathbf{r}'/r$:

$$|\mathbf{r}-\mathbf{r}'| \simeq r - \mathbf{n}\cdot\mathbf{r}'.$$

Then

$$\psi \sim e^{ik\mathbf{n}_0\cdot\mathbf{r}} - \frac{e^{ikr}}{r}\frac{1}{4\pi}\int e^{-ik\mathbf{n}\cdot\mathbf{r}'}V\psi d\mathbf{r}',$$

from which follows

$$A = -\frac{1}{4\pi}\int e^{-ik\mathbf{n}\cdot\mathbf{r}'}V\psi d\mathbf{r}'. \tag{1.6}$$

Let us consider the potentials with which one has to deal in various applications of scattering theory. Potentials which fall off rather abruptly outside a certain radius R, like the screened coulomb potential $1/r \exp(-r/R)$, can be characterized by their range.

In contrast, the potentials decreasing like r^{-n} when $r \to \infty$ cannot be characterized by any finite range. They are often called long-range potentials. Among the long-range potentials of special importance for electronic collisions are: the coulomb potential ($n = 1$), the electric dipole potential ($n = 2$), the quadrupole potential ($n = 3$) and the interaction of a point charge with the induced dipole moment ($n = 4$).

It can be deduced from (1.6) that in the cases $n = 2$ and 3 the amplitude tends to infinity when the scattering angle tends to zero. The total cross section will be finite for $n \geqslant 3$ and infinite for $n < 3$. The case $n = 1$ should be treated separately because (1.2) is invalid and should be modified. The amplitude in this case also tends to infinity when the scattering angle tends to zero.

In the case of scattering by a potential of finite range R, a dimensionless parameter kR can be constructed. If $kR \ll 1$ we will call the particle slow, if $kR \gg 1$ we will call it fast.

The amplitude A has two general properties. One is the reciprocity relation

$$A(\mathbf{n},\mathbf{n}_0) = A(-\mathbf{n}_0, -\mathbf{n}). \tag{1.7}$$

It follows from the consideration of the time-reversed Schrödinger equation in which t is replaced by -t. Another is the so-called optical theorem which is a consequence of the conservation of the norm of the wave function which means that the integral $\int\psi(\mathbf{k},\mathbf{r})\psi^*\cdot(\mathbf{k}',\mathbf{r})d\mathbf{r}$ is the same in the presence of V and in the absence of it. In fact this integral in the case of continuum spectrum wave functions is infinite, but it does not matter. We can consider an integral

$$\int[\psi(k\mathbf{n}_0,\mathbf{r})\psi^*(k\mathbf{n}_0',\mathbf{r}) - e^{ik\mathbf{n}_0\cdot\mathbf{r}}e^{-ik\mathbf{n}_0'\cdot\mathbf{r}}]d\mathbf{r}$$

which is finite. The conservation of the norm means that this expression is zero. The physical meaning of the norm conservation is rather obvious: the potential leads to the redistribution of the particles in a beam over various directions, but it does not change the total number of particles.

Because the most important contribution to the normalization integral comes from large distances, we can use the asymptotic form of the wave function (1.2) and write

$$\int [\psi(k\mathbf{n}_0,\mathbf{r})\psi^*(k\mathbf{n}_0',\mathbf{r}) - e^{ik\mathbf{n}_0\cdot\mathbf{r}}e^{-ik\mathbf{n}_0'\cdot\mathbf{r}}]d\mathbf{r}$$

$$= \int A(\mathbf{n},\mathbf{n}_0)e^{-ik\mathbf{n}_0'\cdot\mathbf{r}}e^{ikr}\frac{d\mathbf{r}}{r} + \int A^*(\mathbf{n},\mathbf{n}_0')e^{ik\mathbf{n}_0\cdot\mathbf{r}}e^{ikr}\frac{d\mathbf{r}}{r} \quad (1.8)$$

$$+ \int A^*(\mathbf{n},\mathbf{n}_0')A(\mathbf{n},\mathbf{n}_0)\frac{d\mathbf{r}}{r^2} = 0.$$

The integration over the absolute value of r begins from some sufficiently large value. The result is

$$\frac{2\pi}{ik}[A^*(\mathbf{n}_0,\mathbf{n}_0') - A(\mathbf{n}_0',\mathbf{n}_0)] + \int A(\mathbf{n},\mathbf{n}_0)A^*(\mathbf{n},\mathbf{n}_0')d\Omega = 0. \quad (1.9)$$

Putting $\mathbf{n} = \mathbf{n}_0'$ here, we obtain the optical theorem

$$\frac{4\pi}{k}\,\mathrm{Im}\,A(\mathbf{n}_0,\mathbf{n}_0) = \int |A(\mathbf{n},\mathbf{n}_0)|^2 d\Omega = \sigma. \quad (1.10)$$

It was mentioned above that the case of a coulomb potential should be treated separately because the asymptotic expression (1.2) is invalid. Now we will consider this problem. The Schrödinger equation for ψ has in this case an exact solution in terms of the confluent hypergeometric functions, so the asymptotical behavior of ψ can be studied in detail. Let us consider the equation for ψ.

In atomic units

$$V = \frac{2ZZ'}{r} \quad (1.11)$$

where Z and Z' are charges of the target and projectile. For electrons Z' = -1. The equation for ψ has the form

$$\Delta\psi + (k^2 + \frac{2Z}{r})\psi = 0. \quad (1.12)$$

To find the solution which in the limit $Z \to 0$ became a plane wave $\exp(i\mathbf{k}\cdot\mathbf{r})$, we put

$$\psi = e^{ikz}f \quad (1.13)$$

where $z = r\cos\theta$, θ is the angle between $\mathbf{k}$ and $\mathbf{r}$. For f we get

$$\Delta f + 2ik\frac{df}{dz} - \frac{2Z}{r}f = 0. \quad (1.14)$$

Let us introduce a new variable $\eta = r - z$. Then

$$\eta\frac{d^2f}{d\eta^2} + \frac{df}{d\eta}(1 - ik\eta) + Zf = 0. \quad (1.15)$$

Comparing (1.15) with the standard form of the equation for the confluent hypergeometric function $F(\alpha,\beta,x)$:

$$x\frac{d^2F}{dx^2} + (\beta - x)\frac{dF}{dx} - \alpha F = 0, \quad (1.16)$$

we see that the solution of (1.15) has the form

$$f = CF(-i\frac{Z}{k},1,ik\eta).$$

Moreover, it should be finite at r = 0. If we specify F as a regular solution of (1.16) which is equal to 1 at x = 0, then C should be chosen as

$$C = \exp(-\frac{\pi Z}{2k})\Gamma(1-i\frac{Z}{k})$$

(the reason for this special choice will become evident further on). The asymptotic form of ψ will be

$$\psi_{r\to\infty} \sim [1+\frac{Z^2}{2ik^3r\sin^2(\frac{\theta}{2})}]\exp\{i[kr\cos\theta-\frac{2Z}{k}\ln(2kr\sin\frac{\theta}{2})]\}$$

$$+ A_c\frac{1}{r}\exp[i(kr+\frac{Z}{k}\ln 2kr)] \qquad (1.17)$$

where

$$A_c = \frac{Z}{2k^2\sin^2(\frac{\theta}{2})}\exp[2i(\frac{Z}{k}\ln\sin\frac{\theta}{2}+\eta_0)],$$

$$\eta_0 = \arg\Gamma(1-i\frac{Z}{k}). \qquad (1.18)$$

It is easy to see that in the limit Z = 0, ψ reduces to the plane wave of unit amplitude. The scattering cross section in the solid angle $d\Omega$ is determined by $|A_c|^2$ and it is equal to the classical Rutherford expression

$$\frac{d\sigma}{d\Omega} = \frac{Z^2}{4k^4\sin^4(\frac{\theta}{2})}. \qquad (1.19)$$

The amplitude A_c tends to infinity at $\theta \to 0$. Therefore the relation (1.10) is not applicable in this case.

1.2 PARTIAL WAVES EXPANSION

Equation (1.1) is rather inconvenient for the direct analysis and calculation of the scattering amplitude. We will transform (1.1) to another form which will provide us with more suitable basis for numerical calculations and general analysis. Let us notice first that there is a partial wave solution of (1.1) in the case of a central field:

$$\psi = \frac{R_\ell(r)}{r}Y_{\ell m}(n). \qquad (2.1)$$

Here $Y_{\ell m}$ is the spherical harmonic which describes a state with given angular momentum and its projection on a fixed direction. The radial function R satisfies the ordinary differential equation

$$\frac{d^2R_\ell}{dr^2} + (k^2 - \frac{\ell(\ell+1)}{r^2} - V)R_\ell = 0. \tag{2.2}$$

The function ψ should be finite at $r = 0$. It follows that $R_\ell(0) = 0$. Let us begin with the case $\ell = 0$. Then

$$\frac{d^2R_0}{dr^2} + (k^2 - V)R_0 = 0. \tag{2.3}$$

If $V = 0$ then the two independent solutions will be e^{ikr} and e^{-ikr}. The solution which vanishes at the origin is $(e^{ikr} - e^{-ikr})$.

To fix the solution completely we will fix the derivative at the origin:

$$(\frac{dR_0}{dr})_{r=0} = 1. \tag{2.4}$$

Then

$$R_0 = \frac{1}{2ik}(e^{ikr} - e^{-ikr}). \tag{2.5}$$

If $V \neq 0$, the solution of (2.3) cannot be written explicitly in analytical form. But it is possible to write down the asymptotic expression in the limit $r \to \infty$ where V vanishes. Taking into account the fact that R_0 as defined by (2.3) and (2.4) should be real, we can write

$$R_0 \sim \frac{1}{2ik}[I_0^*(k)e^{ikr} - I_0(k)e^{-ikr}]. \tag{2.6}$$

The function $I_0(k)$ is called the Jost function. Let us consider some of its general properties. Note that the variable k does not enter in (2.4) and enters in equation (2.3) only in second power. It follows that $R_0(k,r) = R_0(-k,r)$ and, as a result,

$$-\frac{1}{2ik}[I_0^*(-k)e^{-ikr} - I_0(-k)e^{ikr}] = \frac{1}{2ik}[I_0^*(k)e^{ikr} - I_0(k)e^{-ikr}].$$

So we have

$$I_0(-k) = I_0^*(k). \tag{2.7}$$

Using (2.7) we can rewrite (2.6) in the form

$$R_0|_{r\to\infty} \sim \frac{1}{2ik}[I_0(-k)e^{ikr} - I_0(k)e^{-ikr}]. \tag{2.8}$$

Now $I_0(k)$ is in fact a complex function and it can be represented in the form

$$I_0(k) = |I_0(k)|e^{-i\delta_0(k)}. \tag{2.9}$$

The minus sign in the exponent is written for convenience. Then (2.8) is equal to

$$R_0 \sim |I_0(k)| \frac{1}{k} \sin(kr + \delta_0). \tag{2.10}$$

It follows from (2.7) and (2.9) that

$$\delta_0(k) = -\delta_0(-k). \tag{2.11}$$

Let us turn now to the general case $\ell \neq 0$. If $V = 0$ then the regular solution of (2.2) can be expressed in terms of Bessel functions

$$R_\ell = \text{const} \sqrt{\frac{\pi r}{2k}} J_{\ell+\frac{1}{2}}(kr). \tag{2.12}$$

To fix the solution completely we put the coefficient equal to unity. Following the standard notation, we introduce the spherical Bessel functions j_ℓ:

$$j_\ell(x) = \sqrt{\frac{\pi}{2x}} J_{\ell+\frac{1}{2}}(x). \tag{2.13}$$

So in the case $V = 0$, the solution under consideration is

$$R_\ell = r j_\ell(kr). \tag{2.14}$$

The function $j(kr)$ in the limit $r \to 0$ behaves like

$$j_\ell(kr) \sim \frac{(kr)^\ell}{(2\ell + 1)!!} \tag{2.15}$$

and in the limit $r \to \infty$, like

$$j_\ell(kr) \sim \frac{1}{2ikr} [e^{i(kr - \frac{\ell\pi}{2})} - e^{-i(kr - \frac{\ell\pi}{2})}]. \tag{2.16}$$

Let us consider now the general case $V \neq 0$. We will require that in the limit $r \to 0$, the solution of (2.2) behaves still as $rj_\ell(kr)$. That means

$$R_\ell(0) = 0; \quad \frac{d}{dr} (R_\ell r^{-\ell})|_{r=0} = \frac{k^\ell}{(2\ell + 1)!!} .$$

The asymptotic expression of R_ℓ at large r can be written in the form

$$R_\ell \sim \frac{1}{2ik} [I_\ell(-k)e^{i(kr - \frac{\ell\pi}{2})} - I_\ell(k)e^{-i(kr - \frac{\ell\pi}{2})}]. \tag{2.17}$$

Here $I_\ell(k)$ is the Jost function for the angular momentum ℓ. Similar to I_0,

$$I^*_\ell(k) = I_\ell(-k). \tag{2.18}$$

The function $I_\ell(k)$ can be represented in the form similar to (2.9)

$$I_\ell(k) = |I_\ell(k)| e^{-i\delta_\ell(k)}. \tag{2.19}$$

It follows from (2.18) that $\delta_\ell(k) = -\delta_\ell(-k)$, so

$$e^{2i\delta_\ell} = \frac{I_\ell(-k)}{I_\ell(k)} \tag{2.20}$$

With the help of (2.19) the asymptotic expression for R_ℓ can be rewritten in the form

$$R_\ell \sim |I_\ell(k)| \frac{1}{k} \sin(kr - \frac{\ell\pi}{2} + \delta_\ell). \tag{2.21}$$

Let us introduce now the functions f_ℓ according to the relation

$$f_\ell = \frac{1}{I_\ell(k)} R_\ell(kr). \tag{2.22}$$

They satisfy the equation

$$\frac{d^2 f_\ell}{dr^2} + [k^2 - \frac{\ell(\ell+1)}{r^2} - V] f_\ell = 0,$$

vanish at the origin and have at large r the asymptotic form

$$f_\ell \sim e^{i\delta_\ell} \frac{1}{k} \sin(kr - \frac{\ell\pi}{2} + \delta_\ell). \tag{2.23}$$

They are called partial waves. Using the spectroscopic notations, we will denote the waves for $\ell = 0,1,2,3$ by the letters s, p, d, f. Taking into account (2.16), one can rewrite (2.23) in the form

$$f_\ell \sim r j_\ell(kr) + \frac{e^{2i\delta_\ell} - 1}{2ik} e^{i(kr - \frac{\ell\pi}{2})}. \tag{2.24}$$

Now, the product of the partial wave and the spherical harmonic satisfies the equation (1.1), but it certainly does not have the required asymptotic form (1.2). We will try to satisfy (1.2) by the sum over all ℓ of such products multipled by some coefficients.

It is instructive to look at the following expansion of the plane wave

$$e^{i\mathbf{k}\cdot\mathbf{r}} = 4\pi \sum_{\ell,m} i^\ell j_\ell(kr) Y^*_{\ell m}(\mathbf{n}) Y_{\ell m}(\mathbf{n}_o). \tag{2.25}$$

Let us write in a similar fashion

$$\psi = 4\pi \sum_{\ell,m} i^\ell \frac{f_\ell}{r} Y^*_{\ell m}(\mathbf{n}) Y_{\ell m}(\mathbf{n}_o). \tag{2.26}$$

Inserting (2.24) and taking into account (2.25), we get the asymptotic form of

$$\psi \sim e^{i\mathbf{k}\cdot\mathbf{r}} + \frac{e^{ikr}}{r} 4\pi \sum_{\ell,m} \frac{(e^{2i\delta_\ell} - 1)}{2ik} Y^*_{\ell m}(\mathbf{n}) Y_{\ell m}(\mathbf{n}_o). \tag{2.27}$$

It has just the form (1.2). Moreover, we obtain here the expression for A in terms of δ_ℓ. Comparing (2.27) with (1.2) we see that

$$A(\mathbf{n},\mathbf{n}_o) = \frac{2\pi}{ik}\sum_{\ell,m}(e^{2i\delta_\ell}-1)Y^*_{\ell m}(\mathbf{n})Y_{\ell m}(\mathbf{n}_o). \tag{2.28}$$

Taking into account the relation

$$4\pi\sum_m Y^*_{\ell m}(\mathbf{n})Y_{\ell m}(\mathbf{n}_o) = (2\ell+1)P_\ell(\cos\theta)$$

where θ is the angle between $\mathbf{n}$ and $\mathbf{n}_o$, we have

$$A = \frac{1}{2ik}\sum_\ell(2\ell+1)(e^{2i\delta_\ell}-1)P_\ell(\cos\theta). \tag{2.29}$$

The expression

$$A_\ell = \frac{1}{2ik}(e^{2i\delta_\ell}-1) \tag{2.30}$$

is called the partial wave amplitude. From it two other equivalent expressions can be written:

$$A_\ell = \frac{1}{k\cot\delta_\ell - ik}\ ; \quad A_\ell = -\frac{\mathrm{Im}I_\ell}{kI_\ell}\ . \tag{2.31}$$

Note that the expression (2.28) satisfies the relations (1.7) and (1.10) because of the properties of spherical harmonics. By means of the partial waves expansion we reduce the problem of scattering amplitude calculation to the problem of calculation of the phases. It is important to note that under certain conditions only a limited number of partial waves contribute significantly to the scattering amplitude and the rest can be neglected. The conditions in question are suggested by the correspondence between the angular momentum of the scattered particle in classical mechanics $mv\rho$, where ρ is the impact parameter, and that in quantum mechanics $\ell\hbar$. (For the moment we return to the ordinary system of units.) In order to undergo a significant scattering by a potential of a range R, the impact parameter should not exceed the range $\rho \leqslant R$. From this follows that

$$\ell\hbar \lesssim mvR.$$

But $mv/\hbar$ is equal to k. So we have

$$\ell \lesssim kR. \tag{2.32}$$

Now for slow particles $kR \ll 1$. It follows that a significant contribution to the scattering amplitude for slow particles comes from low ℓ only. This result will be confirmed later in another way.
If the potential is long-range then all partial waves should be taken into account even for slow particles.

In some important cases a suitable approximate expression for the phases exists which allows one to sum the series (2.29).

From (2.29) an expression for the total cross section can be deduced

$$\sigma = \frac{4\pi}{k^2}\sum_{\ell}(2\ell + 1)\sin^2\delta_\ell . \tag{2.33}$$

The phases δ_ℓ can be determined by solving the equations for partial waves. But if we are not interested in the wave function itself, it is advantageous to transform the equation for partial waves to the equation for variable phases $\delta_\ell(r)$ which in the limit $r \to \infty$ gives directly the phase δ_ℓ:

$$\delta_\ell(\infty) = \delta_\ell \tag{2.34}$$

and in the limit $r \to 0$ behaves like

$$\delta_\ell(r) \sim (kr)^{2\ell + 1}. \tag{2.35}$$

The equation for $\delta_\ell(r)$ is (Drukarev, 1949):

$$\frac{d\delta_\ell(r)}{dr} = - Vkr^2[j_\ell(kr)\cos\delta_\ell(r) + (-1)^\ell j_{-\ell-1}(kr)\sin\delta_\ell(r)]^2. \tag{2.36}$$

Another form of variable phase method is to introduce a function $t_\ell(r)$ which approaches $\tan\delta_\ell$ when $r \to \infty$:

$$t_\ell(\infty) = \tan\delta_\ell . \tag{2.37}$$

The equation for $t_\ell(r)$ is

$$\frac{dt_\ell}{dr} = - kr^2V[j_\ell(kr) + (-1)^\ell j_{-\ell-1}(kr)t_\ell(r)]^2. \tag{2.38}$$

The derivation of (2.36) and (2.38) and the references to other publications on the variable phase method can be found in the book of F. Calogero (1967). Equation (2.38) will be used later for the discussion of some features of the scattering by a long-range potential.

An important part of scattering theory is the variational principle. It serves as a basis for a computational method which is called the variational method [see the book by Y. Demkov (1963) for details and references to other papers].

We consider here the simplest case $\ell = 0$ only. Let us consider the equation

$$\frac{d^2F}{dr^2} + (k^2 - V)F = 0 \tag{2.39}$$

with the conditions

$$F(0) = 0, \quad F_{r\to\infty} \sim C \sin(kr + \delta_0). \tag{2.40}$$

The normalization factor C will be left at this moment unspecified.

Let us introduce also some function F_t which satisfies the condition (2.40),

$$F_t(0) = 0, \quad F_{t\,r\to\infty} \sim C_t \sin(kr + \delta_t) \tag{2.41}$$

but otherwise is arbitrary and does not satisfy (2.39). This function is called the trial function.

Denote $F_t - F = g$ and consider the integral

$$J_t = \int_0^\infty F_t L F_t dr \tag{2.42}$$

where

$$L = \frac{d^2}{dr^2} + k^2 - V. \tag{2.43}$$

The integral (2.42) is functional with respect to F_t. Taking into account that LF = 0 we can write down an identity

$$J_t = \int_0^\infty FLgdr + \int_0^\infty gLgdr = (Fg' - gF')\Big|_0^\infty + \int_0^\infty gLgdr.$$

Using (2.40) and (2.41), we have

$$(Fg' - gF')\Big|_0^\infty = -kCC_t \sin(\delta_t - \delta_0).$$

It follows that

$$J_t \equiv J_t - J = -CC_t k \sin(\delta_t - \delta_0) + \int_0^\infty gLgdr. \tag{2.44}$$

It is an exact relation. Let us now assume that the difference between δ_t and δ_0 is small. Then up to the second order in $\delta_t - \delta_0$, we have

$$\boldsymbol{\delta} J = -C^2 k \boldsymbol{\delta}\delta_0 \tag{2.45}$$

where the boldface $\boldsymbol{\delta}$ denotes the variation. By specifying C, we can obtain variational principles for different quantities. Putting C = const, we have

$$\boldsymbol{\delta}(J + kC^2\delta_0) = 0. \tag{2.46}$$

Putting $C = 1/k \cos\delta_o$, we have

$$\delta\,(J + \frac{1}{k} \tan\delta_o) = 0 \tag{2.47}$$

which is known as the Kohn principle.

In the variational method which is based on the Kohn principle a trial function is constructed which depends on the parameters $c_1 \ldots c_n$ and has the asymptotic form

$$F_t \sim \frac{1}{k} \sin kr - a_t \cos kr \tag{2.48}$$

where $a_t = -1/k \tan\delta_t$. Then J_t is a function of all the parameters:

$$J_t = J_t(c_1 \ldots c_n, a_t).$$

The stationary condition (2.47) leads to a system of equations

$$\frac{\partial J_t}{\partial c_i} = 0, \quad \frac{\partial J_t}{\partial a_t} = 1. \tag{2.49}$$

Then up to the second order we have

$$(\frac{1}{k} \tan\delta_o)_{Kohn} = St[J_t + \frac{1}{k} \tan\delta_t] \tag{2.50}$$

where by St we denote the stationary value.

This stationary value is not necessarily a minimum. It is a very important point especially if we try to investigate the stability of the value in question under a large variation of parameters.

The various cases which can happen in the simple case of a certain quantity depending upon a single variable parameter are shown on Figure 1.1. The values of $\tan\delta_o$ calculated by the Kohn method can give results similar to curve 3 in Figure 1.1.

However, the case of zero energy is an exception in this respect. Let us consider (2.44) in the limit $k \to 0$. Putting $C = 1/k \cos\delta_o$ and introducing the scattering length a by the relation

$$a = -\lim_{k \to 0} \frac{1}{k} \tan\delta_o,$$

we have

$$J_t - \int_0^\infty gLg dr = a_t - a. \tag{2.51}$$

It was shown by L. Spruch and L. Rosenberg (1959) that

$$\int_0^\infty gLg dr \leqslant 0$$

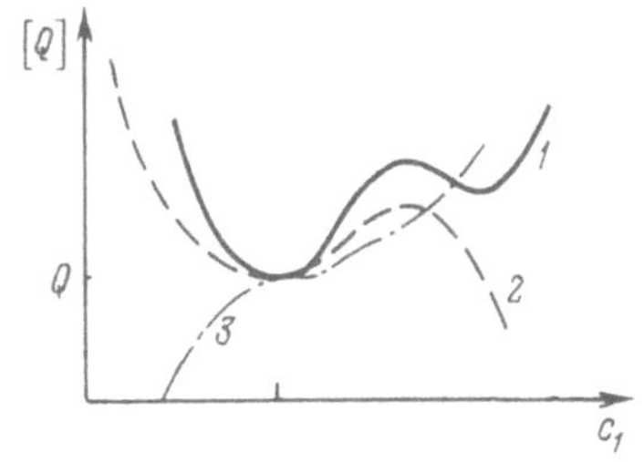

Fig. 1.1 Various types of stability:
1. Stability against large variations;
2. Stability against small variations, instability against large variations;
3. Instability against small variations (Spruch, 1967).

if in the trial function F_t we include the bound state wave functions (provided there are such states). Then

$$a \leqslant a_t - J_t.$$

It follows that

$$\mathrm{St}(a_t - J_t) \geqslant a,$$

but $\mathrm{St}(a_t - J_t)$ is by definition a_{Kohn}. That means that the stationary value of the scattering length is always above the true value.

1.3 THE JOST FUNCTION AND ITS PROPERTIES

In this section some properties of the Jost function used below will be listed.

They all can be derived from the expression

$$I_\ell(k) = 1 + \int_0^\infty h_\ell^{(1)}(kr)V(r)R_\ell(r)r\,dr$$

which can be found in text books on scattering theory (see, for example, Taylor, 1972). Here $h_\ell^{(1)}(kr)$ is related to the Hankel function $H_{\ell+\frac{1}{2}}^{(1)}$ by

$$h_\ell^{(1)}(kr) = \sqrt{\frac{\pi}{2kr}}\, H_{\ell+\frac{1}{2}}^{(1)}(kr).$$

The Jost function can be decomposed into real and imaginary parts in the form

$$I_\ell(k) = C_\ell(k^2) + ik^{2\ell+1}B_\ell(k^2) \qquad (3.1)$$

where C_ℓ and B_ℓ are certain real functions of k^2. If the potential V decreases faster than any power function r^{-n} when $r \to \infty$, then C_ℓ and B_ℓ can be expanded in powers of k^2, $C_\ell(0)$ and $B_\ell(0)$ being finite and in general nonzero. However, if the potential decreases just like the power function r^{-n} when $r \to \infty$, then C_ℓ and B_ℓ will contain functions like $\ln k^2$ which cannot be expanded in powers of k^2.

The Jost function cannot vanish at any real k (except k = 0 in certain special cases which will be discussed below). But the Jost function can have a zero value at certain complex k. According to (2.31) the zero of $I_\ell(k)$ is a pole of the scattering amplitude.

Let us discuss the physical meaning of the zeros of the Jost function. Consider the expression (2.17) for the function R_ℓ at large distances. Assume first that at some point $k = i\kappa$ in the upper half of the complex k-plane, the Jost function is zero:

$$I_\ell(i\kappa) = 0. \tag{3.2}$$

Then the asymptotic expression for R_ℓ will be

$$R_\ell \sim e^{-\kappa r} \tag{3.3}$$

just as it should be in the case of a bound state with the binding energy $\kappa^2/2$.

Since any bound state should have a real energy one can expect that zeros of the Jost function in the upper half of the complex k-plane should be pure imaginary (without a real part). It can be proved rigorously that this is the case indeed. Moreover, it can be proved that there is a relation between the number N_ℓ of bound states with given ℓ and the value of the phase δ_ℓ at $k = 0$:

$$\delta_\ell(0) = N_\ell \pi \tag{3.4}$$

(the so-called Levinson theorem).

There is a special case when one of the zeros of the Jost function is just at the point $\kappa = 0$. In this case (3.4) should be modified. But we will not consider such a special case here.

Let us turn now to the zeros in the lower half of the complex k-plane. Even if the Jost function is equal to zero at certain complex k, the wave function R_ℓ according to (2.17) will increase exponentially when r goes to infinity. It follows that the zeros in the lower half plane have no relation to the bound states. The zeros in the lower half plane can be both pure imaginary and complex. In the later case because of the relation

$$I_\ell(-k^*) = I_\ell^*(k) \tag{3.5}$$

to each zero $k_1 = \eta - i\kappa$ corresponds also a zero at the point

$$k_2 = -\eta - i\kappa.$$

So the complex zeros occur pairwise symmetrically with respect to the imaginary axis.

The physical interpretation of the complex zeros, especially with small imaginary part, can be given in terms of quasi-stationary states. Let us consider a potential which has the form of a well surrounded by a barrier. Assume that a particle is localized inside the well at some initial moment of time which will be taken as zero. The wave function of the particle (which we will suppose to be spherical symmetrical) will be zero outside the barrier at $t = 0$ (Figure 1.2). This state is not a stationary one. Therefore it does not correspond to any definite energy, but instead can be characterized by an energy distribution $W(E)dE$ which is the probability of the energy being between E and $E + dE$. At any time $t > 0$ one can find with certain probability the particle outside the barrier and, as a result, the probability of the particle still being inside the barrier will diminish. There is a close relation between the probability amplitude $L(t)$ of finding at the moment t the particle still inside the barrier and the energy distribution w, which was first established by Fock and Krylov (1947):

$$L(t) = \int_0^\infty e^{-i\frac{Et}{\hbar}} W(E)dE \tag{3.6}$$

(the original derivation of this relation was reproduced in the paper of G. Drukarev, N. Fröman, and P. O. Fröman, 1979). The corresponding probability $P(t)$ is equal to $|L(t)|^2$. The energy distribution W in turn is related to the Jost function. In the case under consideration

$$W = \frac{N}{|I_0|^2} . \tag{3.7}$$

The factor N depends on details of the initial wave function (the derivation of (3.7) and the explicit expression for the factor N can be found in the above cited paper by G. Drukarev, N. Fröman, and P. O.

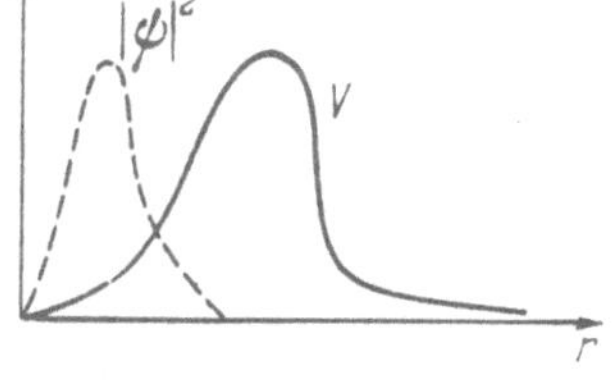

Fig. 1.2 Initial state of a particle inside the barrier.

Fröman, 1979). In the case when there is just one pair of zeros of the Jost function

$$k_1 = \eta - i\kappa, \quad k_2 = -\eta - i\kappa$$

which are near the real axis and all others are far away, it is advantageous to represent I_0 in the form

$$I_0 = U(k)(k - k_1)(k - k_2) \tag{3.8}$$

Inserting (3.8) into (3.7) and introducing the notations

$$E_0 = \frac{1}{2}(\eta^2 + \kappa^2), \quad \Gamma = 2\kappa k, \tag{3.9}$$

we have

$$W = \frac{D(E)}{(E - E_0)^2 + \Gamma^2/4}$$

where in D(E) we absorb everything except the denominator. Now, if $\Gamma \ll E_0$ then W has a form of a sharp peak (Figure 1.3). Neglecting a small shift of the order $(\Gamma/E_0)^2$, the peak is located at the energy E_0. In this approximation κ^2 should be neglected in comparison with η^2, so $E_0 \simeq \eta^2/2$. Through this peak the function D(E) varies but slowly and it can be replaced by a constant.

Due to the normalization condition

$$\int_0^\infty W dE = 1,$$

this constant should be equal to $\Gamma/2\pi$. So we will write finally

$$W(E) = \frac{\Gamma}{2\pi}\frac{1}{(E - E_0)^2 + \Gamma^2/4} \tag{3.10}$$

It is a good approximation provided there is only one pair of zeros of the Jost function near the real axis ($\Gamma \ll E_0$) and all others are

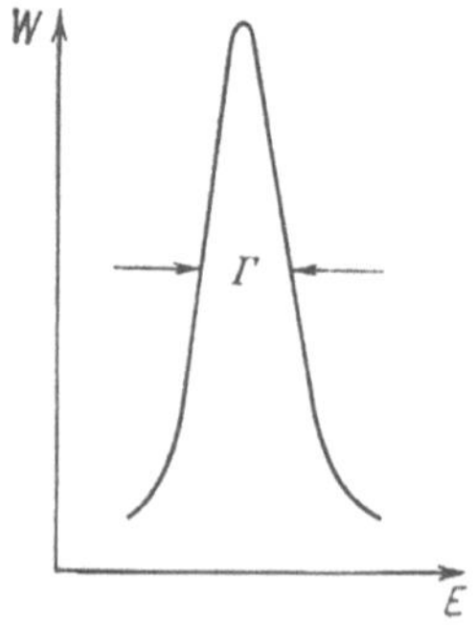

Fig. 1.3 Energy distribution in quasi-stationary state.

far away. Inserting (3.10) into (3.6) and neglecting the quantities of the order $(\Gamma/E_o)^2$, we have

$$L(t) = e^{-i \frac{E_o t}{\hbar} - \frac{\Gamma t}{2\hbar}}$$

and for the probability

$$P(t) = e^{-\frac{\Gamma t}{\hbar}}.$$

It is interesting to note that, according to (2.22), the probability of the penetration of a particle during the scattering process through the barrier is also determined by the factor $|I_\ell|^{-2}$.

It is very instructive to study the motion of the Jost function zeros when the potential is changed (Nussenzweig, 1959; Demkov and Drukarev, 1965). Consider a short range potential well and a particle in an S-state under the action of this potential. Let us select a pair of the complex zeros in the lower half plane (Figure 1.4) and follow their motion when the depth of the well is changed. If the depth of the well is increased the zeros will move towards each other and at certain depth they will coalesce into one double zero at a point c. Increasing the depth further we will see one zero moving upward along the imaginary axis (point a), another moving downward (point b). Crossing the real axis the first zero can move in the upper half plane which corresponds to the formation of a bound state. Bearing in mind this possiblility, the zero of type a is called a virtual state. The zero of type b never will be in the upper half plane and it is called an antibound state. The position of the point c depends on the shape of the potential.

If there is a barrier surrounding the well, then the point c can be quite close to k = 0. In this case there can be zeros with small imaginary part corresponding to long lived quasi-stationary

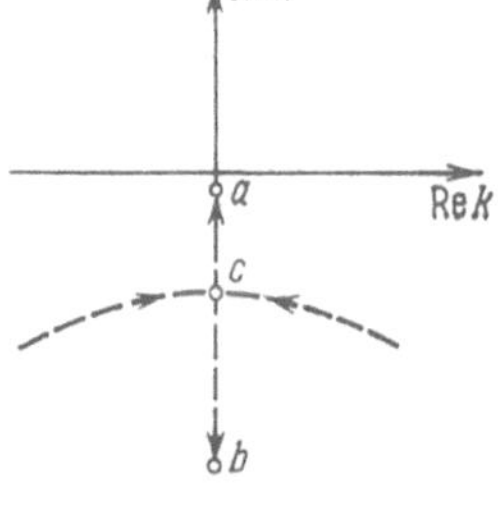

Fig. 1.4 Trajectories of a pair of Jost function zeros near the coalescence point, $\ell = 0$.

states. The point c will move towards $k = 0$ if the penetrability of the barrier decreases.

If there is no barrier then there will be in general no zeros with small imaginary part. It should be noted that the position of complex zeros with large imaginary part is very sensitive to a minor change of the potential.

Let us consider now a state with $\ell \neq 0$ and again select two complex zeros of the Jost function: $k_1 = \eta - i\kappa$, $k_2 = -\eta - i\kappa$. If we increase the depth of the well, k_1 and k_2 will move towards each other and coalesce just at the point $k = 0$ (Figure 1.5), which is easy to understand. There is a centrifugal barrier $\ell(\ell+1)/r^2$ in the case $\ell \neq 0$ and it is impenetrable for $k = 0$. To get an idea about the line along which k_1 and k_2 approach the coalescence point $k = 0$, let us consider $I_\ell(k)$ at small k.

In the expression (3.1) we expand $C_\ell(k^2)$ in powers of k^2 and retain only the first two members: $C_\ell(k^2) = a + bk^2$. The coefficient $B_\ell(k^2)$ we will replace by its value at $k = 0$, $B_\ell(0) = c$. Then

$$I_\ell = a + bk^2 + ick^{2\ell+1}. \tag{3.11}$$

Now we put $k = \pm\eta - i\kappa$ into (3.11) and require that $I_\ell = 0$. Note that in the term $ik^{2\ell+1}$, we should insert $k = \pm\eta$ and drop the imaginary part $i\kappa$, otherwise we would produce real terms smaller than that which we already dropped in the expansion of $C_\ell(k^2)$. Then

$$a + b(\eta^2 - \kappa^2) \pm i\eta(c\eta^{2\ell} - 2b\kappa) = 0.$$

It follows that

$$\eta^2 - \kappa^2 = -a/b, \tag{3.12}$$

$$\eta^{2\ell}/\kappa = {}^{2b}/c. \tag{3.13}$$

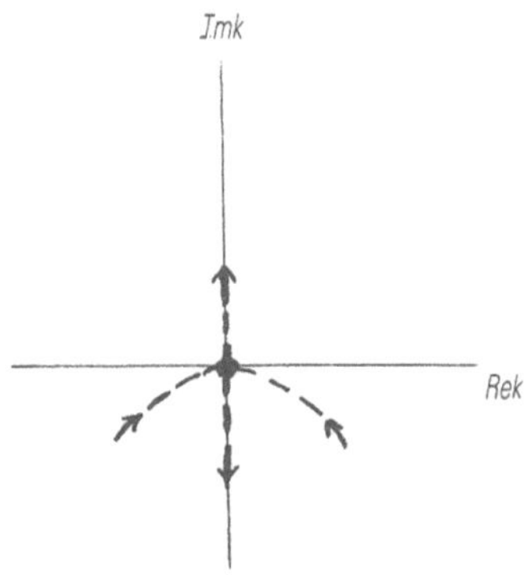

Fig. 1.5 Trajectories of a pair of Jost function zeros near the coalescence point $\ell \neq 0$.

The equations (3.12) and (3.13) express η and κ in terms of a/b and b/c. Any change of potential leads to a change in the values a, b, c and through this to a change in η and κ. It follows from (3.13) that the trajectory of a zero of the Jost function is

$$\kappa = \text{const}\ \eta^{2\ell}. \tag{3.14}$$

After coalescence at k = 0, one of the zeroes moves up the imaginary axis, the other down the imaginary axis.

Their distances from the point k = 0 are almost equal (up to a small correction term). In the case $\ell \neq 0$ it is impossible to have after coalescence one zero near k = 0 and another far away, as in the case $\ell = 0$.

1.4 SCATTERING OF A SLOW PARTICLE BY A SHORT RANGE CENTRAL FIELD: RESONANCES

Inserting (3.1) into (2.20) gives

$$k^{2\ell+1}\cot\delta_\ell = -C_\ell(k^2)/B_\ell(k^2). \tag{4.1}$$

The partial wave amplitude will be equal to

$$A_\ell = -\frac{B_\ell k^{2\ell}}{C_\ell + ik^{2\ell+1}B_\ell}\,.$$

In the case of a short range potential the functions $C_\ell(k^2)$ and $B_\ell(k^2)$ are finite at k = 0. Then one can deduce the behavior of the partial wave amplitude in the limit $k \to 0$. Only A_0 will be finite:

$$A_0(0) = -a \tag{4.2}$$

where a is called the scattering length and is equal to

$$a = -\lim_{k\to 0}\frac{1}{k}\tan\delta_0 = \frac{B_0(0)}{C_0(0)}\,. \tag{4.3}$$

The partial wave amplitude for $\ell \neq 0$ will vanish like $k^{2\ell}$ when $k \to 0$.

Let us consider the scattering length in the important case when V has the shape of a potential well. Putting

$$V = -\lambda V_0 \tag{4.4}$$

where V_0 is positive and has a fixed shape, we will study a as a function of the parameter λ. When $\lambda = 0$ then obviously a = 0. At a certain value of λ the first bound state just appears. At this moment C_0 should be zero because the Jost function is zero when

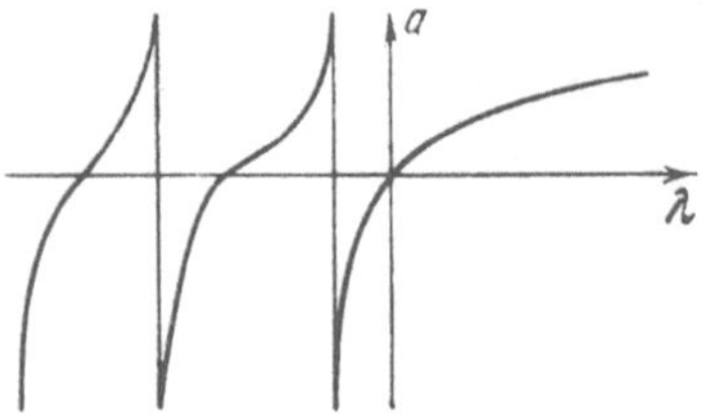

Fig. 1.6 Dependence of scattering length on potential strength.

there is a bound state. According to (4.3) a will be infinite. It will become infinite again when the second bound state just appears, etc. At the intermediate values of λ the function $a(\lambda)$ will behave as shown in Figure 1.6. An interesting example of such a behavior is the scattering of an electron by the Thomas-Fermi atom (see Robinson, 1960). The dependence of a upon the nuclear charge is shown in Fig. 1.7.

We pass now to the consideration of the energy dependence of the scattering amplitude A. Two cases are of special importance:

(a) An imaginary zero of the Jost function very close to the real axis. This corresponds to a bound state with low binding energy (positive imaginary zero) or to a virtual state (negative imaginary zero).
(b) A pair of complex zeros of the Jost function very close to the real axis. This corresponds to a quasi-stationary state.

There is a specific resonance behavior of the cross section as a function of energy in both these cases. In case (a) we put

$$I_0(k) = N(k)(1 - \frac{k}{i\kappa}). \tag{4.5}$$

Here we separate the factor $(1 - k/i\kappa)$ which corresponds to the imaginary zero under consideration. The factor N accumulates all

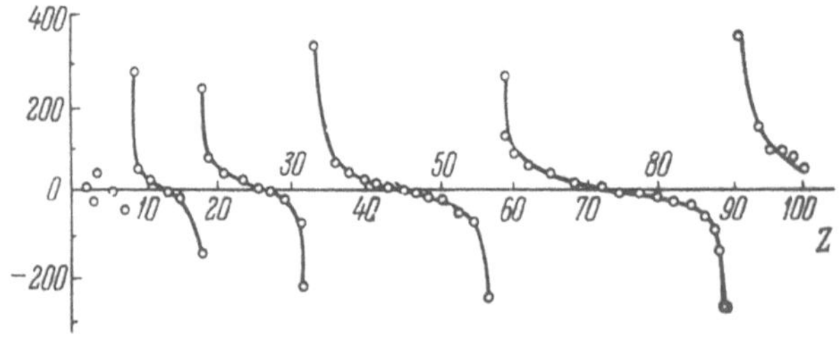

Fig. 1.7 Dependence of the scattering length on Z for the Thomas-Fermi potential (Robinson, 1960)(a in units $a_0 \frac{0.855}{Z}$).

other zeros which are supposed to lie far away. Using (2.20) we put

$$\delta_0 = \delta_0' + \delta_0'', \tag{4.6}$$

$$\exp(2i\delta_0') = \frac{N(-k)}{N(k)}, \tag{4.7}$$

$$\exp(2i\delta_0'') = \frac{i\kappa + k}{i\kappa - k}. \tag{4.8}$$

Here δ_0'' is called the resonant part of the phase δ_0, and δ_0', the background part.

Let us consider the expression $k\cot\delta_0$. Using (4.6), and taking into account (4.8), we have

$$k\cot\delta_0 = \frac{\kappa\, k\cot\delta_0' + k^2}{k\cot\delta_0' + \kappa}. \tag{4.9}$$

Supposing that

$$\kappa \ll k\cot\delta_0'. \tag{4.10}$$

then we get from (4.9) an approximate expression

$$k\cot\delta_0 = -\kappa - \frac{k^2 + \kappa^2}{k\cot\delta_0'}. \tag{4.11}$$

We cannot relate directly the quantity $k\cot\delta_0'$ to some characteristic feature of the potential well V. However, it is possible to do it in a roundabout way. Consider a rectangular potential well of depth V_0 and radius r_0 which can support a bound state with a small binding energy $\kappa^2/2$. Then it can be shown (Landau and Lifshitz) that

$$k\cot\delta_0 = -\kappa + \frac{k^2 + \kappa^2}{2} r_0, \tag{4.12}$$

provided that

$$\kappa r_0 \ll 1. \tag{4.13}$$

Comparing (4.12) with (4.11) we see that

$$k\cot\delta_0' = \frac{2}{r_0}, \tag{4.14}$$

and condition (4.10) is just (4.13).

Now for a potential well of an arbitrary shape we will still use (4.12) and call the parameter r_0 the effective range. By definition

$$\kappa - \frac{1}{2}\kappa^2 r_0 = \frac{1}{a}, \tag{4.15}$$

so (4.12) can be written in the form

$$k\cot\delta_0 = -\frac{1}{a} + \frac{1}{2} r_0 k^2. \tag{4.16}$$

The expression (4.16) was first obtained by Bethe (1949) in another way and is known as the effective range approximation.

It is important to note that this approximation does not depend upon the actual shape of the potential well, provided the two parameters r_0 and a are fixed.

For the partial cross section σ we get

$$\sigma = \frac{4\pi}{(-1/a + k^2 r_0/2)^2 + k^2} \tag{4.17}$$

In the limit $r_0 \to 0$ taking into account (4.15), we obtain the familiar Wigner expression

$$\sigma = \frac{4\pi}{\kappa^2 + k^2}\,. \tag{4.18}$$

Owing to the condition (4.13) the cross section near zero energy will be much larger than the "geometrical" one $4\pi r_0^2$.

The limit $r_0 \to 0$ corresponds to the zero range potential model in which the Schrödinger equation inside the potential well is replaced by the boundary condition imposed on the wave function at the origin r = 0:

$$\left[\frac{1}{r\psi}\frac{d}{dr}(r\psi)\right]_{r=0} = -\frac{1}{a} \tag{4.19}$$

(Bethe and Peierls, 1935; Drukarev, 1978; Demkov and Ostrovsky, 1975). If we represent ψ in the form

$$\psi = e^{i\mathbf{k}\cdot\mathbf{r}} + A\,\frac{e^{ikr}}{r} \tag{4.20}$$

and insert this expression in (4.19), then for the scattering amplitude we will have

$$A = -\frac{1}{1/a + ik} \tag{4.21}$$

which leads to the expression (4.18) for σ.

In case (b) we put instead of (4.5)

$$I_0(k) = N(k)(1 - k/k_1)(1 - k/k_2) \tag{4.22}$$

where

$$k_1 = \eta - i\kappa, \quad k_2 = -\eta - i\kappa. \tag{4.23}$$

Again we divide the phase δ_0 into the resonant and background parts

$$\delta_0 = \delta_0' + \delta_0''$$

$$\exp(2i\delta_0') = \frac{N(-k)}{N(k)}$$

$$\exp(2i\delta_0'') = \frac{(k_1+k)(k_2+k)}{(k_1-k)(k_2-k)} \,. \tag{4.24}$$

Inserting (4.23) in (4.24) and denoting

$$E_0 = \frac{1}{2}(\eta^2+\kappa^2)$$
$$\Gamma = 2\kappa k, \tag{4.25}$$

we have

$$\cot\delta_0'' = \frac{2(E_0 - E)}{\Gamma} \tag{4.26}$$

In this case we will not try to relate $\cot\delta_0'$ to any feature of the potential and merely put

$$\cot\delta_0' = q. \tag{4.27}$$

Denoting also

$$\cot\delta_0'' = \varepsilon$$

we get for $\cot\delta_0$, an expression

$$\cot\delta_0 = \frac{q\varepsilon - 1}{q+\varepsilon} \,, \tag{4.28}$$

and for the cross section

$$\sigma = \sigma' \,\frac{(q+\varepsilon)^2}{1+\varepsilon^2} \tag{4.29}$$

where σ' is the background cross section

$$\sigma' = \frac{4\pi}{k^2}\sin^2\delta'. \tag{4.30}$$

The expression for the cross section (4.29) was derived by U. Fano.

Let us consider (4.29) in detail. In Figure 1.8 the dependence of the ratio σ/σ' upon ε is shown for various q, assuming $q > 0$. The same curves can be used for $q < 0$ if we put on the abscissa $-\varepsilon$ instead of ε. Strictly speaking q is not a fixed parameter but is rather a function of the energy E. But because $\varepsilon = 2(E_0 - E)/\Gamma$ and $\Gamma \ll E_0$, a very small change of E near the point E_0 corresponds to a large change on the ε scale. It follows that we can use the fixed value $q(E_0)$ on the ε scale provided ε is not too large. For any given value of q the ratio σ/σ' will be zero at the point $\varepsilon = -q$ and will reach its maximum at the point $\varepsilon = 1/q$. The value at the maximum will be

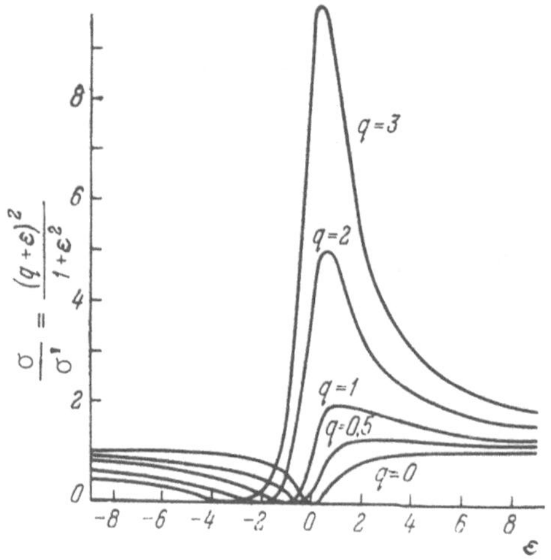

Fig. 1.8 σ/σ' plot for various q (Schulz, 1973).

$$\left(\frac{\sigma}{\sigma'}\right)_{max} = 1 + q^2. \tag{4.31}$$

Taking into account (4.3) we have

$$\sigma_{max} = \frac{4\pi}{k^2}. \tag{4.32}$$

In the limit $\delta' \to 0$ (or $q \to \infty$) which corresponds to neglecting the background scattering, we have

$$\sigma = \frac{4\pi}{k^2}\frac{1}{1+\varepsilon^2} \tag{4.33}$$

which in view of (4.26) reduces to the well-known Breit-Wigner formula

$$\sigma = \frac{\pi}{k^2}\frac{\Gamma^2}{(E_0 - E)^2 + 1/4\Gamma^2}\ . \tag{4.34}$$

It should be noted that the expression (4.33) has a maximum at the point $\varepsilon = 0$ (or $E = E_0$) only on the ε scale. Let us see what happens on the energy scale. Having in mind that $E_0 = (\eta^2 + \kappa^2)/2$, $\Gamma = 2k\kappa$, one can transform (4.34) to the expression

$$\sigma = \frac{\pi}{\eta^2}\frac{\gamma^2}{(E_0 - E - \gamma^2/4\eta^2) \ + \gamma^2/4} \tag{4.35}$$

where

$$\gamma = 2\kappa\eta. \tag{4.36}$$

It is easy to see that the maximum value of σ will be reached at the point

$$E_{max} = E_0 - \gamma^2/4\eta^2 = E_0 - \kappa^2. \tag{4.37}$$

The position of the maximum is shifted from the point E_0 toward lower energy.

If $\kappa \ll \eta$, the ratio $\gamma^2/E_0\eta^2$ is very small and the shift can be neglected. But if η and κ are of the same order of magnitude then the shift can be large enough and there will be no maximum at all at positive energy E.

Let us now turn to the case $\ell \neq 0$. We consider first scattering by a rectangular potential well of a radius r_0. If the well can hold just one bound state with a given ℓ having very small binding energy w then (Drukarev, 1981)

$$k^{2\ell+1}\cot\delta_\ell = -p_\ell \frac{E+w}{r_0^{2\ell+1}} \tag{4.38}$$

where

$$p_\ell = \frac{2\ell+1}{2\ell-1}[(2\ell-1)!!]^2$$

If the well cannot hold any bound state with this ℓ and there is a quasi-stationary state instead with a small energy E_0, then

$$k^{2\ell+1}\cot\delta_\ell = -p_\ell \frac{E-E_0}{r_0^{2\ell-1}} . \tag{4.39}$$

For a potential well of arbitrary shape we will still use (4.38) and (4.39) treating r_0 as an effective range. Then (4.38) and (4.39) represent the effective range approximation for $\ell \neq 0$. Finally, let us express the width of the resonance Γ in terms of the effective range r_0. Comparing (4.39) with the expression for $\cot\delta_\ell$ in the vicinity of resonance

$$\cot\delta_\ell = \frac{2(E_0-E)}{\Gamma}$$

and putting at the resonance $k^{2\ell+1} \simeq \eta^{2\ell-1}$, we have

$$\Gamma = \frac{2}{p_\ell}\eta^{2\ell+1} r_0^{2\ell-1}. \tag{4.40}$$

1.5 EFFECT OF A LONG-RANGE TAIL IN THE POTENTIAL WHICH FALLS OFF AS r^{-4}

Let us consider the effect of a polarization potential which falls off at large distances from the origin as

$$V_p \sim -\alpha/r^4 \tag{5.1}$$

(see Section 0.2). We consider first the case $\ell = 0$. Let us surround the target by a sphere of a radius R large enough to neglect all short-range interactions in the region $r > R$ and to consider V_p as a small perturbation. We will assume also that $kR \ll 1$. To determine $\tan\delta_0$ in the region $r \geqslant R$, we use the equations (2.38) and (2.39)

$$\frac{dt_0}{dr} = \frac{\alpha}{kr^4}[\sin kr + t_0(r)\cos kr]^2. \tag{5.2}$$

In the first approximation we replace $t_0(r)$ in the right side of (5.2) by $t_0(R)$, its value being determined by the inner region $r \leqslant R$. Then from (5.2) we have

$$\tan\delta_0 = t_0(R) + \frac{\alpha}{k}\left[\int_R^\infty \frac{\sin^2 kr}{r^4}dr + t_0(R)\int_R^\infty \frac{\sin 2kr}{r^4}dr + t_0^2(R)\int_R^\infty \frac{\cos^2 kr}{r^4}dr\right]. \tag{5.3}$$

In evaluating the integrals entering in the right side of (5.3) we neglect all terms containing $1/R$ and take into account that $kR \ll 1$. Then the integrals $\int_R^\infty r^{-4}\sin^2 kr\,dr$ and $\int_R^\infty r^{-4}\cos^2 kr\,dr$ are reduced to the expression $k^3\int_0^\infty x^{-4}(x^2 - \sin^2 x)dx$. Using the relation

$$\int_0^\infty x^{-4}(x^2 - \sin^2 x)dx = \frac{\pi}{3},$$

we find that

$$\int_R^\infty \frac{\sin^2 kr}{r^4}dr \simeq -\frac{\pi}{3}k^3, \tag{5.4}$$

$$\int_R^\infty \frac{\cos^2 kr}{r^4}dr \simeq \frac{\pi}{3}k^3. \tag{5.5}$$

The integral

$$\int_R^\infty \frac{\sin 2kr}{r^4}dr$$

is reduced to the cosine integral $\int_R^\infty r^{-4}\sin 2kr\,dr \simeq \frac{4}{3}k^3\mathrm{Ci}(2kR)$. Using the asymptotic expression for Ci(z) for $z \ll 1$ we find that

$$\int_R^\infty r^{-4}\sin 2kr\,dr \simeq \frac{4}{3}k^3(\ln 2kr + \gamma) \tag{5.6}$$

where γ is the Euler constant.

Now let us consider $t_0(R)$. We expand it in powers of k retaining only the first two terms:

$$t_0(R) = c_1 k + c_2 k^3.$$

The coefficient c_1 is related to the scattering length

$$c_1 = -a$$

because in the approximation used here,

$$-a = \lim_{k\to 0} \frac{1}{k} \tan\delta_0 = \lim_{k\to 0} \frac{1}{k} t_0(R).$$

Taking this into account, we write $t_0(R)$ in the form

$$t_0(R) = -ak + c_2k^3. \tag{5.7}$$

The next term in (5.7) would be $\sim k^5$, but we will drop this term (as well as all others of higher order).

We insert the expression (5.7) into (5.3). We should drop all terms containing k^5 and higher order terms because they are dropped in (5.7). Then

$$\tan\delta_0 = -ak - \frac{\pi\alpha}{3}k^2 + bk^3 - \frac{4}{3}ak^3\ln k + \frac{\pi a^2\alpha}{3}k^4 \tag{5.8}$$

where b accumulates all numerical coefficients in front of k^3:

$$b = c_2 - \frac{4}{3}a(\ln 2R + \gamma - 1).$$

In applications of (5.8) the coefficient b should be considered as an adjustable parameter. This expression is often called the modified effective range expansion. An expression similar to (5.8) but without the term $\pi a^2\alpha k^4/3$ was obtained and used by O'Malley et al., (1961), and O'Malley (1963).

From (5.8) some important features of the partial cross section σ_0 can be deduced. First let us consider the slope of the curve $\sigma_0(k)$ at $k = 0$. In the vicinity of $k = 0$ the cross section σ_0 will be

$$\sigma = 4\pi\left(a + \frac{\pi\alpha}{3}k + \ldots\right)^2. \tag{5.9}$$

We dropped all terms of higher order than k because they do not contribute to $(d\sigma_0/dk)_{k=0}$. From (5.9) we have

$$\left(\frac{d\sigma_0}{dk}\right)_{k=0} = \frac{8\pi^2\alpha a}{3}. \tag{5.10}$$

In the absence of the long-range interaction $(d\sigma_0/dk)_{k=0}$ would be zero. As a result of the long-range tail $V_p(r)$ it is different from zero. The sign of $(d\sigma_0/dk)_{k=0}$ is determined by the sign of a. When $a < 0$ then $(d\sigma_0/dk)_{k=0} < 0$. The cross section decreases and at a certain value of k will become zero. This is the Ramsauer effect (Figure 1.9). The critical value at which $\sigma_0 = 0$ can be estimated from (5.9)

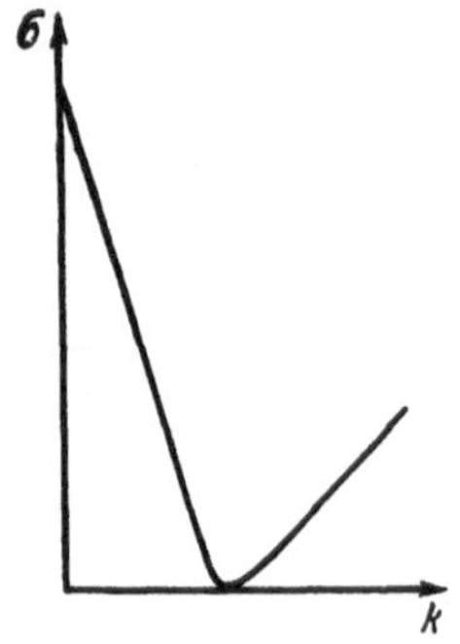

Fig. 1.9 Ramsauer effect for the $\ell = 0$ partial cross section.

$$k_o \simeq -\frac{3a}{\pi\alpha} \tag{5.11}$$

(up to correction terms of higher order in k). It should be noted that it is only the $\ell = 0$ partial cross section which vanishes at $k = k_o$. The actual observed cross section will have a deep minimum rather than a true zero. When $a > 0$ then $(d\sigma_o/dk)_{k=0} > 0$ and there will be a maximum in the partial cross section σ_o.

Now we pass to the case $\ell \neq 0$. The simplest approximation will be to drop in Equation (2.38) the term $t_\ell(r)(-1)^\ell j_{-\ell-1}$ and to write

$$\frac{dt_\ell}{dr} = \frac{\alpha k}{r^2}\, j_\ell^2(kr). \tag{5.12}$$

It follows that

$$\tan\delta_\ell = t_\ell(R) + \alpha k\int_R^\infty \frac{dr}{r^2}\, j_\ell^2(kr).$$

Taking into account (2.13), we get

$$\tan\delta_\ell = t_\ell(R) + \frac{\pi\alpha k^2}{2}\int_{kR}^\infty J^2_{\ell+\frac{1}{2}}(x)\,\frac{dx}{x^3}\,.$$

Using the condition that $kR \ll 1$, we can replace the lower limit of the integral kR by zero. Using the relation

$$\int_0^\infty J^2_{\ell+\frac{1}{2}}(x)\,\frac{dx}{x^3} = \frac{1}{4(\ell+3/2)(\ell+\frac{1}{2})(\ell-\frac{1}{2})}\,,$$

we get

$$\tan\delta_\ell = t_\ell(R) + \frac{\pi\alpha k^2}{8(\ell+3/2)(\ell+\frac{1}{2})(\ell-\frac{1}{2})}\,. \tag{5.13}$$

This result shows a radical change of the energy dependence of $tg\delta_\ell$ in comparison with a short range potential. The polarization potential leads to one and the same dependence $\sim k^2$ for all ℓ while any short range potential leads to a $k^{2\ell+1}$ dependence. For all ℓ except $\ell = 1$, it is quite sufficient at small k to drop $t_\ell(R)$ and retain only the k^2 term. For $\ell = 1$ we will retain also $t_0(R)$ which is proportional to k^3.

Now we are able to calculate the scattering amplitude for small k. Replacing in the first approximation $\exp(2i\delta_\ell) - 1$ by $2i\delta_\ell$ for all $\ell \geq 1$ and separating the $\ell = 0$ term, we have

$$A = A_0 + \frac{1}{k} \sum_{\ell=1}^{\infty} (2\ell + 1)\delta_\ell P_\ell(\cos\theta). \tag{5.14}$$

In the same approximation we replace $\tan\delta_\ell$ by δ_ℓ and insert in (5.14). Inserting here the expression (5.13), dropping $t_\ell(R)$ for all $\ell > 1$ and retaining the contribution $t_1(R)$, we get

$$A = A_0 + \frac{\pi\alpha k}{4} \sum_{\ell=1}^{\infty} \frac{P_\ell(\cos\theta)}{(\ell + 3/2)(\ell - \frac{1}{2})} + ck^2\cos\theta. \tag{5.15}$$

The sum

$$\sum_{\ell=1}^{\infty} \frac{P_\ell(\cos\theta)}{(\ell + 3/2)(\ell - \frac{1}{2})}$$

can be evaluated. The result is

$$\sum_{\ell=1}^{\infty} \frac{P_\ell(\cos\theta)}{(\ell + 3/2)(\ell - \frac{1}{2})} = \frac{4}{3} - 2\sin\frac{\theta}{2}. \tag{5.16}$$

It follows that

$$A = A_0 + \frac{\pi\alpha k}{3} - \frac{\pi\alpha k}{2}\sin\frac{\theta}{2} + ck^2\cos\theta. \tag{5.17}$$

Note the unusual term $\sin\theta/2$ which cannot appear in the amplitude for any short range potential.

With the help of (5.17) we can calculate the differential cross section $d\sigma/d\Omega = |A|^2$. It is interesting to consider $(d|A|^2/d\theta)_{\theta=0}$. Using (5.17) we can deduce that

$$\left(\frac{d|A|^2}{d\theta}\right)_{\theta=0} = -\frac{\pi\alpha k}{4}\left(A_0 + \frac{\pi\alpha k}{3} + ck^2\right). \tag{5.18}$$

The sign of the right hand side of (5.18) depends upon the sign of the sum

$$A_0 + \pi\alpha k/3 + ck^2.$$

In the limit $k \to 0$ according to (5.8)

$$A_0 + \frac{\pi\alpha k}{3} \to -a.$$

As a result for small k the sign of $(d|A|^2/d\theta)_{\theta=0}$ will be the same as the sign of scattering length a. But if the energy is increased, then the sign of $A_0 + \pi\alpha k/3 + ck^2$ can change and, as a result, the sign of $(d|A|^2/d\theta)_{\theta=0}$ will change.

Finally, we note that the long-range tail r^{-4} in the potential alters significantly not only the cross section, but also the behavior of the wave function. Consider the limiting case of zero energy. The asymptotic form of the wave function (1.2) reduces to the expression

$$\psi \sim 1 - \frac{a}{r} . \tag{5.19}$$

The difference between the exact wave function and its asymptotic form

$$|\psi - (1 - a/r)| \tag{5.20}$$

will decrease when $r \to \infty$.

How fast it decreases depends upon the potential. If the potential is short range and falls off like $\exp(-\alpha r)$, then the expression (5.20) will decrease also like $\exp(-\alpha r)$. But for a long-range potential $\sim r^{-4}$ the expression (5.20) decreases like r^{-2}.

1.6 EFFECT OF THE COULOMB ATTRACTIVE POTENTIAL ON SLOW PARTICLE SCATTERING

Another case of the influence of a long-range tail in a potential on the scattering which is important in atomic physics is the Coulomb attractive potential superimposed on a short range potential. The scattering of an electron on a positive atomic ion is an example.

Let us represent the potential in the form

$$V = \frac{-2Z}{r} + V_1 \tag{6.1}$$

and the scattering amplitude in the form

$$A = A_c + A' . \tag{6.2}$$

A_c is the Coulomb amplitude given by (1.18). The partial amplitude A_ℓ also will be the sum

$$A_\ell = A_{c\ell} + A'_\ell . \tag{6.3}$$

The partial Coulomb amplitude

$$A_{c\ell} = \frac{1}{2ik}(e^{2i\eta_\ell} - 1)$$

where

$$\exp(2i\eta_\ell) = \frac{\Gamma(\ell+1-i\,Z/k)}{\Gamma(\ell+1+i\,Z/k)}\;.$$

If we represent A_ℓ in the standard form

$$A_\ell = \frac{1}{2ik}(e^{2i\delta_\ell}-1)$$

and put

$$\delta_\ell = \eta_\ell + \mu_\ell,$$

then for A'_ℓ we will get

$$A'_\ell = \frac{1}{2ik}\,e^{2i\eta_\ell}(e^{2i\mu_\ell}-1). \tag{6.4}$$

The problem is reduced to the calculation of the phase μ_ℓ due to the short range part of potential in the presence of the Coulomb tail.

To calculate μ_ℓ we will use an equation like (2.36) but replace rj_ℓ and $rj_{-\ell-1}$ by the regular and irregular radial Coulomb wave functions (Babikov, 1968)

$$\frac{d\mu_\ell}{dr} = -\frac{V_1}{k}\,[F_\ell\cos\mu_\ell(r) + G_\ell\sin\mu_\ell(r)]^2. \tag{6.5}$$

In the limit $k \to 0$ for an attractive Coulomb potential, all the dependence of F_ℓ and G_ℓ upon k concentrates in a factor $k^{1/2}$. The rest will be independent of k as $k \to 0$.

It follows that in the limit $k \to 0$ equation (6.5) will not contain k at all. The correction terms at small but finite k will be of order k^2. It follows then that at small k the phase μ_ℓ can be expressed as

$$\mu_\ell(k) = \mu_\ell(0) + \frac{1}{2}\,\mu''_\ell(0)k^2 + \ldots \tag{6.6}$$

Besides the scattering states with positive energy there also can be bound states in the potential (6.1).

In the absence of the short range part in the potential V the energy of a bound state is given by the well-known expression

$$E_n = -\,Z^2/2n^2. \tag{6.7}$$

In the presence of V_1 it is customary to write

$$E_{n\ell} = -\,Z^2/2(n-\Delta_\ell)^2 \tag{6.8}$$

where Δ_ℓ is the so-called quantum defect.

There is an interesting and important relation between μ_ℓ and Δ_ℓ (Seaton, 1958):

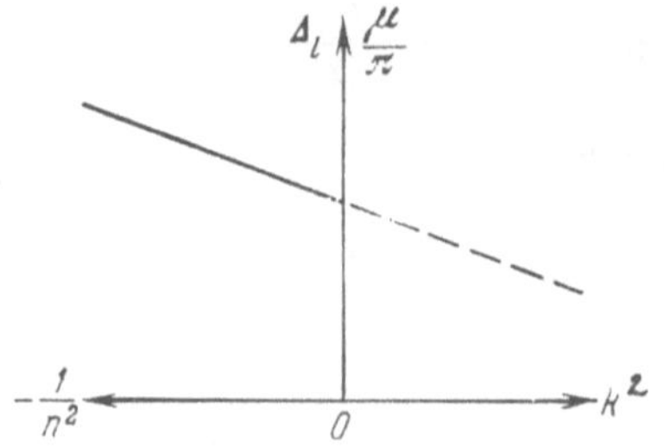

Fig. 1.10 Quantum defect extrapolation (Seaton, 1958).

$$\lim_{n\to\infty} \pi\Delta_\ell(n) = \mu_\ell(0). \tag{6.9}$$

At large n the quantum defect depends almost linearly upon n^{-2}. We can extrapolate this dependence outside the negative energy part of the energy spectra and find $\mu_\ell(k^2)$ (Figure 1.10).

It should be noted that an addition of an integral multiple of π to the phase μ_ℓ has no effect on the scattering amplitude. It follows that only the fractional part of Δ_ℓ matters.

Now let us turn to the part A' of the scattering amplitude A in (6.2). Taking into account (6.4), we have

$$A' = \frac{1}{2ik}\sum_\ell (2\ell+1)e^{2i\eta_\ell}(e^{2i\mu_\ell}-1)P_\ell(\cos\theta).$$

It is convenient to represent A' in the form

$$A' = \frac{1}{k}\, e^{2i\eta_o} f. \tag{6.10}$$

Using the expression for η_ℓ and the known properties of Γ-functions, one can deduce that

$$\exp 2i(\eta_\ell - \eta_o) = \frac{(\ell - i\, Z/k)\ \ldots\ (1 - i\, Z/k)}{(\ell + i\, Z/k)\ \ldots\ (1 + i\, Z/k)}\,.$$

It follows that in the limit $k \to 0$ the function f will be a function of the scattering angle θ only.

For further use we will write down

$$f = |f|e^{i\alpha}. \tag{6.11}$$

Finally, we get for the scattering amplitude A the following:

$$A = e^{2i\eta_o}\left\{\frac{Z\exp[i\, Z/k\, \ln \sin^2(\theta/2)]}{2k^2\sin^2(\theta/2)} + \frac{e^{i\alpha}|f|}{k}\right\} \tag{6.12}$$

and for the differential cross section

$$\frac{d\sigma}{d\Omega} = \frac{Z^2}{4k^4 \sin^4(\theta/2)} + \frac{|f|^2}{k^2} + \frac{Z|f|}{k^3 \sin^2(\theta/2)} \cos[\frac{Z}{k} \ln \sin^2(\theta/2) + \alpha]. \tag{6.13}$$

Let us fix some value of the scattering angle θ and consider $d\sigma/d\Omega$ as a function of k.

If we increase k starting from $k = 0$ then first the term $\cos(\frac{Z}{k} \ln \sin^2 \frac{\theta}{2} + \alpha)$ will be an oscillating function of k, but after some critical value of k there will be no oscillations.

This critical value depends upon θ. It decreases when θ is increasing. At $\theta = \pi$ there will be no oscillations at all. In this case

$$(\frac{d\sigma}{d\Omega})_{\theta=\pi} = \frac{Z^2}{4k^4} + \frac{|f|^2}{k^2} + \frac{Z}{k^3} |f| \cos\alpha . \tag{6.14}$$

Taking into account that $|f|\cos\alpha$ is in fact Re f, we write (6.14) in the form

$$(\frac{d\sigma}{d\Omega})_{\theta=\pi} = \frac{Z^2}{4k^4} + \frac{|f|^2}{k^2} + \frac{Z}{k^3} \operatorname{Re} f . \tag{6.15}$$

Depending on the sign and value of Re f, the cross section can be larger than Coulomb, or smaller.

At certain particular values of $|f|$ and Re f, there can be a deep minimum in the cross section as a function of k. It is an analog of the Ramsauer effect.

The oscillations of the cross section can be seen also at fixed k by changing θ.

It follows from (6.13) that the oscillations are concentrated in the region of small θ, when $\ln \sin^2(\theta/2)$ is large enough. At angles near π this logarithmic term is very small and there will be no oscillations.

It should be noted that in general many phases μ_ℓ contributed to the function f as follows from the definition of this function and the properties of μ_ℓ.

1.7 SCATTERING BY TWO SHORT-RANGE NONOVERLAPPING POTENTIALS

In this section we will consider the scattering of a slow particle by two short-range nonoverlapping potentials. This problem is important in connection with electron molecule scattering (see Section 8.2).

To make this problem tractable we will replace the actual short range potential by a zero-range potential model. This means that boundary conditions like (4.19) will be used at each center instead of the Schrödinger equation.

Let us put the origin of the coordinate system just in the middle between the two centers so that their position will be equal to $-\mathbf{R}/2$ and $\mathbf{R}/2$, R being the distance between the centers.

We represent the wave function of the scattered particles in the form

$$\psi = e^{ik\mathbf{r}\cdot\mathbf{n}_0} + A_1 \frac{e^{ik|\mathbf{r}-\mathbf{R}/2|}}{|\mathbf{r}-\mathbf{R}/2|} + A_2 \frac{e^{ik|\mathbf{r}+\mathbf{R}/2|}}{|\mathbf{r}+\mathbf{R}/2|} . \tag{7.1}$$

In the limit $\mathbf{r} \to \mathbf{R}/2$

$$\psi \sim e^{ik\mathbf{n}_0\cdot\mathbf{R}/2} + A_1(ik + \frac{1}{|\mathbf{r}-\mathbf{R}/2|}) + A_2 \frac{e^{ikR}}{R} . \tag{7.2}$$

The boundary condition at the center $\mathbf{R}/2$ will be

$$[\frac{1}{\rho_1\psi} \frac{d}{d\rho_1} (\rho_1\psi)]_{\rho_1\to 0} = -\frac{1}{a_1} \tag{7.3}$$

where

$$\rho_1 = |\mathbf{r}-\mathbf{R}/2| . \tag{7.4}$$

From this boundary condition we get

$$A_1(ik + \frac{1}{a_1}) + A_2 \frac{e^{ikR}}{R} = -e^{ik\mathbf{n}_0\cdot\mathbf{R}/2} . \tag{7.5}$$

Near the second center, $\mathbf{r} \to -\mathbf{R}/2$, we have

$$\psi \sim e^{-ik\mathbf{n}_0\cdot\mathbf{R}/2} + A_1 \frac{e^{ikR}}{R} + A_2(ik + \frac{1}{|\mathbf{r}+\mathbf{R}/2|}) . \tag{7.6}$$

Using the boundary condition

$$[\frac{1}{\rho_2\psi} \frac{d}{d\rho_2} (\rho_2\psi)]_{\rho_2\to 0} = -\frac{1}{a_2} \tag{7.7}$$

where

$$\rho_2 = |\mathbf{r} + \mathbf{R}/2|, \tag{7.8}$$

we get

$$A_1 \frac{e^{ikR}}{R} + A_2(ik + \frac{1}{a_2}) = - e^{-ik\mathbf{n}_0\cdot\mathbf{R}/2}. \tag{7.9}$$

From (7.5) and (7.9) the coefficients A_1 and A_2 can be determined.

The final step is to express the scattering amplitude A in terms of A_1 and A_2. It follows from (7.1) in the limit $r \to \infty$ that

$$\psi \sim e^{ik\mathbf{r}\cdot\mathbf{n}_0} + \frac{e^{ikr}}{r} (A_1 e^{-ik\mathbf{n}\cdot\mathbf{R}/2} + A_2 e^{ik\mathbf{n}\cdot\mathbf{R}/2}). \tag{7.10}$$

By definition

$$A(\mathbf{n},\mathbf{n}_0) = A_1 e^{-ik\mathbf{n}\cdot\mathbf{R}/2} + A_2 e^{ik\mathbf{n}\cdot\mathbf{R}/2}. \tag{7.11}$$

First, let us derive an expression for the scattering length

$$a = - A_{k=0}. \tag{7.12}$$

It follows from (7.11) that

$$a = - (A_1 + A_2)_{k=0}. \tag{7.13}$$

To determine A_1 and A_2, we put $k = 0$ in (7.5) and (7.9). The resulting equations are

$$\frac{1}{a_1} A_1 + \frac{1}{R} A_2 = - 1 \tag{7.14}$$

$$\frac{1}{R} A_1 + \frac{1}{a_2} A_2 = - 1. \tag{7.15}$$

From (7.14), (7.15) and (7.13) we get

$$a = \frac{a_1 + a_2 - 2a_1a_2/R}{1 - a_1a_2/R^2}. \tag{7.16}$$

In Figure 1.11 the shape of the curve a(R) is shown for various combinations of a_1 and a_2. In the case $a_1 = a_2$

$$a = \frac{2a_1}{1 + a_1/R}. \tag{7.17}$$

If $a_1 < 0$ then the curve a(R) will have a shape like Figure 1.11c. If $a_1 > 0$ then the shape will be like Figure 1.11b.

Now, let us consider the amplitude A in the case $a_1 = a_2$. It is equal to

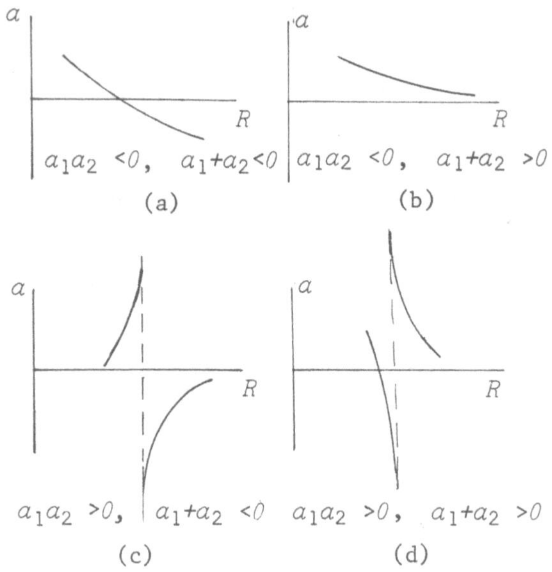

Fig. 1.11 Scattering length as a function of R.

$$A = -2\left[\frac{\cos(k\mathbf{n}\cdot\mathbf{R}/2)\cos(k\mathbf{n}_0\cdot\mathbf{R}/2)}{1/a_1 + ik + \exp(ikR)/R} + \frac{\sin(k\mathbf{n}\cdot\mathbf{R}/2)\sin(k\mathbf{n}_0\cdot\mathbf{R}/2)}{1/a_1 + ik - \exp(ikR)/R}\right]. \tag{7.18}$$

The term $\exp(ikR)/R$ represents the multiple scattering effect. First we notice that in the limit $k \to 0$ one term remains finite, and another vanishes $\sim k^2$.

In other words one term (the first in the brackets) is like scattering in an s state. Another term (the second in the brackets) is like scattering in a p state.

If $1/a_1 + \cos(kR)/R = 0$, then $A \to \infty$ when $k \to 0$. This corresponds to a resonance. It is due to the multiple scattering term $\exp(ikR)/R$.

If we neglect multiple scattering, then

$$A = -\frac{2}{1/a_1 + ik}\cos[k\mathbf{R}\cdot(\mathbf{n} - \mathbf{n}_0)/2]. \tag{7.19}$$

The factor $\cos[k\mathbf{R}\cdot(\mathbf{n} - \mathbf{n}_0)/2]$ is like the interference of two waves coherently emitted by the sources located at $\mathbf{R}/2$ and $-\mathbf{R}/2$.

1.8 SCATTERING BY AN ELECTRIC DIPOLE

The interaction energy of an electron with two point charges of opposite sign Z and -Z is given by the expression

$$U = Z\left(\frac{1}{r_1} - \frac{1}{r_2}\right) \tag{8.1}$$

where r_1 and r_2 are the distances between the electron and the point charges Z and -Z.

The corresponding scattering problem was considered by many authors. We will follow the treatment given by Abramov and Komarov (1975) but reproduce here only the main results omitting various computational details. We define a cartesian coordinate system putting its origin just in the middle point between the charges and drawing the Z-axis through charges. The radius-vector of an electron will be denoted by r. It is advantageous to use prolate spheroidal coordinates

$$\xi = \frac{r_1 + r_2}{R}, \quad \eta = \frac{r_1 - r_2}{R}, \quad \phi = \arctan\frac{y}{x} \tag{8.2}$$

Here R is the distance between the charges Z and - Z. The spheroidal coordinates are limited by the following inequalities:

$$1 \leqslant \xi < \infty, \quad -1 \leqslant \eta \leqslant 1, \quad 0 \leqslant \phi \leqslant 2\pi. \tag{8.3}$$

Let us represent the wave function ψ in the form

$$\psi = X(\xi)Y(\eta)e^{im\phi}. \tag{8.4}$$

Inserting (8.4) and (8.1) into the Schrödinger equation and denoting

$$D = 2RZ, \tag{8.5}$$

we get

$$\begin{aligned} &\frac{d}{d\xi}(\xi^2 - 1)\frac{dX}{d\xi} + \left[\frac{k^2R^2}{4}(\xi^2 - 1) - \frac{m^2}{\xi^2 - 1} - C\right]X = 0, \\ &\frac{d}{d\eta}(1 - \eta^2)\frac{dY}{d\eta} + \left[\frac{k^2R^2}{4}(1 - \eta^2) - \frac{m^2}{1 - \eta^2} + D\eta + C\right]Y = 0. \end{aligned} \tag{8.6}$$

Here C is the so-called "separation constant" which in fact is a function of kR and D rather than a true constant, and has to be determined.

The functions X and Y must be finite at the points ± 1 and 1.

The second of equations (8.6) generates the system of eigenfunctions $Y_{nm}(\eta)$ which are orthogonal. We will impose the usual normalization conditions on the functions Y_{nm}. They are called angular because in the limit $r \to \infty$

$$\lim_{r\to\infty} \eta = \cos\theta$$

where θ is the angle between the unit vector along $\mathbf{r}$ and the z-axis.

The functions X_{nm} which are generated by the first of equations (8.6) are called radial. To determine completely the function X_{nm}, we must fix the asymptotic behavior of X_{nm} in the limit $\xi \to \infty$.

We will assume

$$X_{nm} \sim \frac{1}{\xi kR/2} \sin[\frac{kR}{2}\xi + \Delta_{nm} - \frac{\pi}{2}(n + |m|)] \tag{8.7}$$

and require that Δ_{nm} vanishes if $D = 0$.

The scattering amplitude can be expressed in terms of the angular functions $Y_{nm}(\eta)$

$$A(\mathbf{n},\mathbf{n}_0) = 1/ik \sum_{n,m} (2 - \delta_{om})\cos m(\phi - \phi_0) Y_{nm}(\cos\theta) \times [\exp 2i(\Delta_{nm} - n\pi/2) Y_{nm}(-\cos\theta_0) - Y_{nm}(\cos\theta_0)]. \tag{8.8}$$

Here the angles θ,ϕ and θ_0,ϕ_0 determine the unit vectors $\mathbf{n}$ and $\mathbf{n}_0$.

From (8.8) we can deduce that the amplitude becomes infinite in the limit $\mathbf{n} \to \mathbf{n}_0$ except when $\theta_0 = 0$ and $\theta_0 = \pi$, when it is finite. The total cross section also becomes infinite for any $\mathbf{n}_0$ except the case when $\mathbf{n}_0$ is along the z-axis.

In this case

$$\sigma = \frac{2\pi}{k^2} \sum_{n=0}^{\infty} \{4\sin^2\Delta_{no} Y_{no}(1) Y_{no}(-1)(-1)^n + |Y_{no}(1) - (-1)^n Y_{no}(-1)|^2\}. \tag{8.9}$$

The behavior of the phases Δ_{nm} in the limit $kR \to 0$ can be inferred from the relation

$$\exp 2i[\Delta_{nm} - \frac{\pi}{2}(n + |m| + \tfrac{1}{2})] = \frac{\exp(-i\pi\nu) - \varepsilon}{1 - \varepsilon\exp(-i\pi\nu)} \tag{8.10}$$

where

$$\nu = (C + 1/4)^{\frac{1}{2}} \tag{8.11}$$

and ε for small kR is equal to

$$\varepsilon \simeq (\frac{kR}{8})^{2\nu} \frac{\Gamma^2(1-\nu)}{\Gamma^2(1+\nu)} \frac{\Gamma(\frac{1}{2} + \nu - |m|)}{\Gamma(\frac{1}{2} - \nu - |m|)} . \tag{8.12}$$

In (8.12) we dropped all terms containing higher powers of kR than 2ν.

Looking at (8.11) we notice that there are two cases to consider: real ν, when $C > -1/4$ and imaginary ν when $C < -1/4$. In the first case there is no bound state of an electron in the field of a dipole. In the second case there is a sequence of bound states. If ν is real, then

$$\lim_{kR\to 0} \Delta_{nm} = \Delta_{nm}^{(0)} = \frac{\pi}{2}[n + |m| + \frac{1}{2} - \nu]. \tag{8.13}$$

If ν is imaginary then (8.10) can be reduced to the form

$$\exp 2i[\Delta_{nm} - \frac{\pi}{2}(n + |m| + \frac{1}{2})] = \frac{\exp(\pi|\nu|) - \exp i\Phi}{1 - \exp(\pi|\nu|)\exp i\Phi} \tag{8.14}$$

where

$$\Phi = 2|\nu| \ln kR + \arg \frac{\Gamma^2(1 - i|\nu|)\Gamma(\frac{1}{2} + i|\nu| - |m|)}{\Gamma^2(1 + i|\nu|)\Gamma(\frac{1}{2} - i|\nu| - |m|)} . \tag{8.15}$$

Now Φ increases logarithmically when $kR \to 0$. Then $\exp i\Phi$ will oscillate and so will the phases Δ_{nm} in the limit $kR \to 0$.

When Δ_{nm} approaches the value $\pi/2$ or $3\pi/2$, then there will be a resonance in the cross section.

The zeroes of the denominator in (8.14) at imaginary k give the bound states.

It remains to see how to determine the "separation constant" C. We are interested in small values of kR. Then it is sufficient to study the equation for the angular function in the limit $k = 0$,

$$\frac{d}{d\eta}(1 - \eta^2)\frac{dY}{d\eta} + [C - \frac{m^2}{1 - \eta^2} + D\eta]Y = 0. \tag{8.16}$$

If $D = 0$ then this equation reduces to that for the associated Legendre polynomials $P_{n+|m|}^{|m|}$ and

$$C = (n + |m|)(n + |m| + 1). \tag{8.17}$$

If $D > 0$ then C will be lower than the value given by (8.17).

Increasing D will eventually give a zero value of C. Finally, at sufficiently high D the value of C will become lower than $-1/4$. According to the above discussion, this leads to the appearance of bound states in the field (8.1).

1.9 SCATTERING OF FAST PARTICLES. VARIOUS APPROXIMATIONS

In this section the scattering of a fast particle by a short-range potential of a range R is considered. For the fast particle $kR \gg 1$; we will also assume that the potential energy of the particle U is small in comparison with its total energy E

$$|U|/E \ll 1. \tag{9.1}$$

Then on the average the scattering angle will be small.

Let us discuss the conditions under which the classical treatment of the scattering is valid. In the classical theory the basic relation is the dependence of scattering angle θ upon the impact parameter ρ. It is necessary that the quantum uncertainties $\Delta\theta$ and $\Delta\rho$ should be small in comparison to θ and ρ

$$\Delta\rho \ll \rho, \quad \Delta\theta \ll \theta. \tag{9.2}$$

Now the scattering angle under condition (9.1) can be approximately expressed by the relation

$$\theta = \left| \frac{p_\perp}{p} \right| \ll 1 \tag{9.3}$$

where $p_\perp$ is the component of the particle momentum p, perpendicular to its initial direction.

For $\Delta\theta$ we get

$$\Delta\theta \sim \Delta p_\perp / p. \tag{9.4}$$

According to the Heisenberg uncertainty relation

$$\Delta p_\perp \sim \hbar/\Delta\rho.$$

Then

$$\Delta\theta \sim 1/k\Delta\rho. \tag{9.5}$$

If we require that $\Delta\rho \ll \rho$, and take into account that only $\rho \lesssim R$ matters, then

$$\Delta\theta \gg 1/kR.$$

It follows from (9.2) that

$$\theta \gg 1/kR. \tag{9.6}$$

In the region $\theta \lesssim 1/kR$ the scattering cannot be described by classical mechanics even if $kR \gg 1$. The scattering in this region is analogous to wave diffraction.

Now let us estimate the value of the scattering angle θ in the region where classical mechanics can be used.

The transverse component $p_\perp$ of the momentum which enters in (9.3) can be calculated as

$$p_\perp = \int_{-\infty}^{\infty} F_\perp dt \tag{9.7}$$

where $F_\perp$ is the force acting on the scattered particle. It depends upon the impact parameter ρ.

Bearing in mind that $p_\perp/p$ is small, we will neglect in the integral (9.7) the deviation of the trajectory from a straight line and suppose that the particle moves with constant velocity v along the z axis.

Then $dt = dz/v$ and

$$p_\perp = \frac{1}{v}\int_{-\infty}^{\infty} F_\perp dz. \tag{9.8}$$

Expressing $F_\perp$ in terms of the potential energy $U(\rho,z)$

$$F_\perp = -\frac{\partial U}{\partial \rho},$$

we have

$$p_\perp = -\frac{1}{v}\frac{d}{d\rho}\int_{-\infty}^{\infty} U(\rho,z)dz. \tag{9.9}$$

Let us introduce the function

$$\delta(\rho) = -\frac{1}{2\hbar v}\int_{-\infty}^{\infty} U(\rho,z)dz. \tag{9.10}$$

Then

$$p_\perp = 2\hbar\frac{d\delta}{d\rho},$$

and for the scattering angle θ we get

$$\theta = (2/k)\frac{d\delta}{d\rho}. \tag{9.11}$$

It is important to note that the function $\delta(\rho)$ is just the expression which follows from the semiclassical approximation for the phase δ_ℓ if we expand δ_ℓ in powers of the ratio U/E and retain only the first term, and replace ℓ by $k\rho$. Taking this into account we can rewrite (9.11) in the form

$$\theta = 2\,\frac{d\delta_\ell}{d\ell}\,. \tag{9.12}$$

With the help of (9.11) the condition (9.6) can be expressed as

$$\frac{d\delta(\rho)}{d\rho} >> \frac{1}{R}\,. \tag{9.13}$$

If the condition (9.13) is not satisfied, then the scattering should be considered using quantum mechanics.

Instead of the scattering angle as a function of an impact parameter the scattering amplitude as a function of the angle should be considered.

The relevant approximation was developed by Glauber (1959) and it is called the Glauber approximation. To derive the basic expression for the amplitude, let us represent the wave function of a particle in the form

$$\psi = e^{i\mathbf{k}\cdot\mathbf{r}}\Phi \tag{9.14}$$

and assume that Φ is a slow varying function of the coordinate z (the z axis is taken along $\mathbf{k}$).

Inserting (9.14) into (1.1) and neglecting the second derivative of Φ we get for the function Φ an equation

$$2ik\,\frac{\partial\Phi}{\partial z} = V\Phi, \tag{9.15}$$

with the solution being

$$\Phi = \exp\Big[-\frac{i}{2k}\int_{-\infty}^{z} V(x,y,z)dz\Big]. \tag{9.16}$$

Inserting (9.14) and (9.16) into the expression for the amplitude (1.6), we get

$$A = -\frac{1}{4\pi}\int d\mathbf{r}e^{i\mathbf{q}\cdot\mathbf{r}}V\exp\Big[-\frac{i}{2k}\int_{-\infty}^{z} Vdz\Big] \tag{9.17}$$

where

$$\mathbf{q} = k(\mathbf{n}_0 - \mathbf{n})$$

is the change in the momentum.

At small scattering angles the vector $\mathbf{q}$ is approximately perpendicular to the vector $\mathbf{k}$ (Figure 1.12).

Let us introduce the cylindrical coordinates ρ, z:

$$q_\rho = k\sin\theta, \quad q_z = 2k\sin^2\frac{\theta}{2}\,.$$

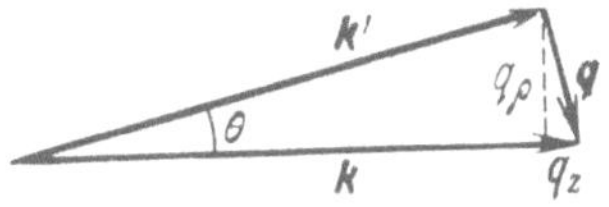

Fig. 1.12 Momentum vector diagram.

It follows from this relation that when $\theta \ll 1$ then $q_z \ll q_\rho$. If we transform the integrand in (9.17) to cylindrical coordinates and neglect q_z, then the integration over z can be performed with the result

$$A = \frac{ik}{2\pi}\int d\boldsymbol{\rho} e^{iq\rho}[1 - e^{2i\delta(\boldsymbol{\rho})}] \tag{9.18}$$

where $\delta(\rho)$ is given by (9.10). The expression (9.18) is the Glauber approximation.

Denoting

$$\Gamma = 1 - e^{2i\delta}, \tag{9.19}$$

we can represent (9.18) in the form

$$A = \frac{ik}{2\pi}\int d\boldsymbol{\rho} e^{iq\rho}\Gamma(\boldsymbol{\rho}). \tag{9.20}$$

If there are several short range centers located at the points ρ_i (Figure 1.13), then

$$\delta = -\frac{1}{2k}\sum_i\int V_i(\boldsymbol{\rho} - \mathbf{s}_i, z)dz = \sum_i \delta_i. \tag{9.21}$$

Accordingly the Γ-factor can be represented in the form

$$\Gamma = 1 - \prod_i e^{2i\delta_i} = 1 - \prod_i(1 - \Gamma_i).$$

It follows that

$$\Gamma = \sum_i \Gamma_i - \sum_{i\neq k}\Gamma_i\Gamma_k + \ldots \tag{9.22}$$

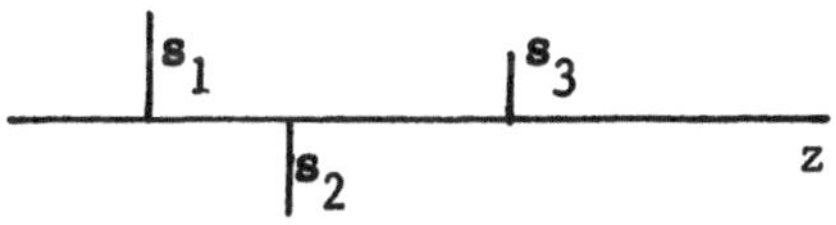

Fig. 1.13

The first term in the right side of (9.22) represents the independent scattering from each center and the other terms represent the effect of rescattering, or multiple scattering of the particle by the system of the centers.

For long range potentials the treatment should be modified. The classical theory for the small angle scattering by a number of Coulomb centers was developed by Demkov (1981) by means of a special complex impact parameter method. The application of the Glauber approximation to the same problem was developed with the help of complex impact parameters by Abramov, Demkov and Shcherbakov (1981).

Finally, let us consider the Born approximation. It follows from (1.6) that if we replace $\psi(\mathbf{r}')$ in the integrand by a plane wave $\exp(ik\mathbf{n}_0\cdot\mathbf{r}')$, then we get for the amplitude the expression

$$A = -\frac{1}{4\pi}\int e^{i\mathbf{q}\cdot\mathbf{r}} V d\mathbf{r}. \tag{9.23}$$

It depends only upon the vector

$$\mathbf{q} = k(\mathbf{n}_0 - \mathbf{n}).$$

For the absolute value of $\mathbf{q}$ there is an expression

$$q = 2k\sin(\theta/2). \tag{9.24}$$

The Born approximation can be obtained also from the Glauber approximation if the phase $\delta(\rho)$ is small and $1 - \exp 2i\delta$ is replaced by the first term of the expansion in powers of δ, that is, $2i\delta$.

The condition $\delta(\rho) \ll 1$ under which the Born approximation is valid is equal to

$$(|\bar{V}|/k)R \ll 1 \tag{9.25}$$

for short range potentials.

In the case where there are several short range centers the amplitude in Born approximation will be equal to the sum of the scattering amplitudes due to each of the centers. Multiple scattering is neglected in this approximation.

2
Scattering of a Particle with Spin: Polarization Phenomena

2.1 AMPLITUDE MATRIX

The theory considered in the previous chapter did not take into account the spin of the particle. We will discuss now how to incorporate spin in the description of the scattering. Since we will be considering electrons we will suppose that the spin of the scattered particle is equal to 1/2 (in the units of $\hbar$).

If we are going to consider atoms as a target, then we should include the target spin also. The simplest case is an atom with zero total spin and therefore we will start with the case of a spinless target.

Let us assume for the time being that all electrons in the incident beam are in the same spin state which is characterized by a given projection of the spin μ on a certain fixed direction. This direction will be taken as the z axis of the coordinate system.

Later we will consider the more realistic case when there is a certain polarization of the electronic beam rather than a state with a given projection.

The wave function of a free electron is

$$\psi_\mu = u_\mu e^{i\mathbf{k}\cdot\mathbf{r}} \tag{1.1}$$

where u_μ is an eigenfunction of the z-projection spin operator

$$\hat{S}_z u_\mu = \mu u_\mu . \tag{1.2}$$

The eigenvalues μ are $\pm$ 1/2 (in the units of $\hbar$). In matrix representation $\hat{S}_{x,y,z}$ are 2 x 2 matrices acting on the 2-component column vector

$$u_{1/2} = \begin{pmatrix} 1 \\ 0 \end{pmatrix}, \quad u_{-1/2} = \begin{pmatrix} 0 \\ 1 \end{pmatrix}. \tag{1.3}$$

It is customary to put

$$\hat{S}_{x,y,z} = \frac{1}{2}\sigma_{x,y,z}$$

where $\sigma_{x,y,z}$ are the Pauli matrices

$$\sigma_x = \begin{pmatrix} 0 & 1 \\ 1 & 0 \end{pmatrix}, \quad \sigma_y = \begin{pmatrix} 0 & -i \\ i & 0 \end{pmatrix}, \quad \sigma_z = \begin{pmatrix} 1 & 0 \\ 0 & -1 \end{pmatrix}. \tag{1.4}$$

We will use Dirac's notation and instead of a matrix column vector $\mathbf{u}_\mu$, we write a spin state vector $|\mu\rangle$.

It follows directly from (1.3) - (1.4) that

$$\sigma_x|1/2\rangle = |-1/2\rangle, \quad \sigma_x|-1/2\rangle = |1/2\rangle$$

$$\sigma_y|1/2\rangle = i|-1/2\rangle, \quad \sigma_y|-1/2\rangle = -i|1/2\rangle$$

$$\sigma_z|1/2\rangle = |1/2\rangle, \quad \sigma_z|-1/2\rangle = -|-1/2\rangle. \tag{1.5}$$

In the presence of a target the asymptotic expression for the electronic wave function can be written in the form

$$\psi = e^{i\mathbf{k}\cdot\mathbf{r}}|\mu\rangle + \frac{e^{ikr}}{r}\sum_{\mu'}|\mu'\rangle\langle\mu'\mathbf{n}|A|\mu\mathbf{n}_o\rangle. \tag{1.6}$$

$$\mu,\mu' = -1/2, 1/2$$

Instead of a single scattering amplitude $A(\mathbf{n},\mathbf{n}_o) \equiv \langle\mathbf{n}|A|\mathbf{n}_o\rangle$, we now have four amplitudes $\langle\mu'\mathbf{n}|A|\mu\mathbf{n}_o\rangle$ which form an amplitude matrix.

The square of the absolute value $|\langle\mu'\mathbf{n}|A|\mu\mathbf{n}_o\rangle|^2$ is the differential cross section for scattering into a unit solid angle around $\mathbf{n}$ accompanied by a transition from an initial spin state μ to a final one μ'. In order to make the expressions more compact we will drop sometimes the indices $\mathbf{n}$ and $\mathbf{n}_o$ (when it will not be misleading) and write

$$\langle\mu'|A|\mu\rangle \text{ instead of } \langle\mu'\mathbf{n}|A|\mu\mathbf{n}_o\rangle.$$

To get an insight into the structure of the amplitude matrix, let us consider an operator acting on the spin variables μ. Having in mind the scattering amplitude, we will require that the operator should be invariant under the rotation of the coordinate system, and under inversion of time and space coordinates.

There are only two independent invariants which can be built in the spin 1/2 case: the unit operator I and a scalar product $\boldsymbol{\sigma}\cdot\boldsymbol{\nu}$ where $\boldsymbol{\nu}$ is some axial vector. Accordingly we will write down the amplitude operator $\hat{A}$ in the form

$$\hat{A} = u(\mathbf{n},\mathbf{n}_0)I + v(\mathbf{n},\mathbf{n}_0)(\boldsymbol{\sigma}\cdot\boldsymbol{\nu}). \tag{1.7}$$

It remains to see what axial vector $\boldsymbol{\nu}$ can be built from the only available vectors $\mathbf{n}_0$ and $\mathbf{n}$. It is indeed the vector product $[\mathbf{n},\mathbf{n}_0]$. If we want $\boldsymbol{\nu}$ to be a unit vector, then

$$\boldsymbol{\nu} = \frac{[\mathbf{n},\mathbf{n}_0]}{|[\mathbf{n},\mathbf{n}_0]|}\ . \tag{1.8}$$

It is normal to the scattering plane $\{\mathbf{n},\mathbf{n}_0\}$.

Now, using (1.5) we have

$$\begin{aligned}
&<\tfrac{1}{2}\ |A|\ \tfrac{1}{2}> = u + v\nu_z \\
&<-\tfrac{1}{2}\ |A|\ -\tfrac{1}{2}> = u - v\nu_z \\
&<-\tfrac{1}{2}\ |A|\ \tfrac{1}{2}> = <\tfrac{1}{2}\ |A|\ -\tfrac{1}{2}>^* = v(\nu_x + i\nu_y).
\end{aligned} \tag{1.9}$$

It follows that the four amplitudes $\langle\mu'|A|\mu\rangle$ are expressed in terms of two functions u and v which are in general complex

$$u = |u|e^{i\alpha}, \quad v = |v|e^{i\beta}. \tag{1.10}$$

If we draw the z axis of the coordinate system along $\boldsymbol{\nu}$ then the amplitude matrix will be diagonal with respect to the spin variables μ,μ'. In other words the spin projection on the normal to the scattering plane remains unchanged. However, the scattering amplitudes corresponding to "spin down" and "spin up" are different. The nature of this difference is the same as the fine structure of atomic energy levels. It is a relativistic effect and it vanishes in the nonrelativistic approximation.

Now we pass to the case of a target having its total electronic spin equal to 1/2, the hydrogen atom being the simplest example.

We will denote the scattered electron spin projection by the index 1, and that of atomic electron by the index 2.

The asymptotic form of the wave function is a direct generalization of (1.6):

$$\psi \sim e^{i\mathbf{k}\cdot\mathbf{r}}|\mu_1\mu_2\rangle + \frac{e^{ikr}}{r}\sum_{\mu_1'\mu_2'}|\mu_1'\mu_2'\rangle\langle\mu_1'\mu_2'\mathbf{n}|A|\mu_1\mu_2\mathbf{n}_0\rangle. \tag{1.11}$$

The amplitude $\langle\mu_1'\mu_2'\mathbf{n}|A|\mu_1\mu_2\mathbf{n}_o\rangle$ describes the transition of the scattered electron's spin from μ_1 to μ_1' and that of the atomic electron from μ_2 to μ_2'.

Again we will consider an operator which acts on the spin variables $\mu_1\mu_2$, using invariants constructed from $\boldsymbol{\sigma}_1$, $\boldsymbol{\sigma}_2$, $\mathbf{n}$ and $\mathbf{n}_o$.

Let us first define the three orthogonal unit vectors

$$\boldsymbol{\lambda} = \frac{\mathbf{n}_o + \mathbf{n}}{|\mathbf{n}_o + \mathbf{n}|}; \quad \boldsymbol{\mu} = \frac{\mathbf{n}_o - \mathbf{n}}{|\mathbf{n}_o - \mathbf{n}|}; \quad \boldsymbol{\nu} = \frac{[\mathbf{n},\mathbf{n}_o]}{|[\mathbf{n},\mathbf{n}_o]|}. \tag{1.12}$$

Here $\boldsymbol{\nu}$ is an axial vector while $\boldsymbol{\lambda}$ and $\boldsymbol{\mu}$ are polar vectors. With the help of $\boldsymbol{\lambda}$, $\boldsymbol{\mu}$, $\boldsymbol{\nu}$ the following independent invariants can then be constructed:

$$(\boldsymbol{\sigma}_1\cdot\boldsymbol{\lambda})(\boldsymbol{\sigma}_2\cdot\boldsymbol{\lambda}), \quad (\boldsymbol{\sigma}_1\cdot\boldsymbol{\mu})(\boldsymbol{\sigma}_2\cdot\boldsymbol{\mu}), \quad (\boldsymbol{\sigma}_1\cdot\boldsymbol{\nu})(\boldsymbol{\sigma}_2\cdot\boldsymbol{\nu}), \quad (\boldsymbol{\sigma}_1\cdot\boldsymbol{\nu}) \text{ and } (\boldsymbol{\sigma}_2\cdot\boldsymbol{\nu}).$$

It follows that the operator $\hat{A}$ can be represented in the form

$$\hat{A} = a_1 + a_2(\boldsymbol{\sigma}_1\cdot\boldsymbol{\lambda})(\boldsymbol{\sigma}_2\cdot\boldsymbol{\lambda}) + a_3(\boldsymbol{\sigma}_1\cdot\boldsymbol{\mu})(\boldsymbol{\sigma}_2\cdot\boldsymbol{\mu}) + a_4(\boldsymbol{\sigma}_1\cdot\boldsymbol{\nu})(\boldsymbol{\sigma}_2\cdot\boldsymbol{\nu}) + a_5(\boldsymbol{\sigma}_1\cdot\boldsymbol{\nu}) + a_6(\boldsymbol{\sigma}_2\cdot\boldsymbol{\nu}). \tag{1.13}$$

Here $a_1 \ldots a_6$ are functions of the unit vectors $\mathbf{n}_o$ and $\mathbf{n}$, like u and v in (1.7).

Note that the matrix elements of an operator like $(\boldsymbol{\sigma}_1\cdot\boldsymbol{\lambda})(\boldsymbol{\sigma}_2\cdot\boldsymbol{\lambda})$ can be decomposed in the following way:

$$\langle\mu_1'\mu_2'|(\boldsymbol{\sigma}_1\cdot\boldsymbol{\lambda})(\boldsymbol{\sigma}_2\cdot\boldsymbol{\lambda})|\mu_1\mu_2\rangle = \lambda_x^2\langle\mu_1'|\sigma_{1x}|\mu_1\rangle\langle\mu_2'|\sigma_{2x}|\mu_2\rangle + \lambda_y^2\langle\mu_1'|\sigma_{1y}|\mu_1\rangle\langle\mu_2'|\sigma_{2y}|\mu_2\rangle + \lambda_z^2\langle\mu_1'|\sigma_{1z}|\mu_1\rangle\langle\mu_2'|\sigma_{2z}|\mu_2\rangle. \tag{1.14}$$

Such a decomposition will be used below in the calculations.

The expression (1.13) takes into account all relativistic effects like the spin-orbit interaction.

If we consider the nonrelativistic limit then the conservation of the total spin projection should be taken into acount

$$\mu_1' + \mu_2' = \mu_1 + \mu_2. \tag{1.15}$$

To ensure this, it is necessary to put

$$a_2 = a_3 = a_4; \quad a_5 = 0; \; a_6 = 0.$$

Then the general expression (1.13) reduces to the form

$$\hat{A} = a_1 + a_2(\boldsymbol{\sigma}_1\cdot\boldsymbol{\sigma}_2). \tag{1.16}$$

Let us study in detail this nonrelativistic limit. The only nonvanishing amplitudes will be

$$\langle\mu\mu|A|\mu\mu\rangle = a_1 + a_2 \tag{1.17}$$

$$\langle\mu,-\mu|A|\mu,-\mu\rangle = a_1 - a_2 \tag{1.18}$$

$$\langle\mu,-\mu|A|-\mu,\mu\rangle = 2a_2 \tag{1.19}$$

where $\mu = \pm 1/2$.

The amplitude $\langle\mu,-\mu|A|-\mu,\mu\rangle$ corresponds to the so-called spin-flip amplitude. In the literature $-2a_2 = g$ is often referred to as the "exchange" amplitude and $a_1 - a_2 = f$ as the "direct" amplitude.

To get further insight into the structure of the amplitude matrix we will consider the scattering process in terms of total spin S_t and its projection μ_t.

In the nonrelativistic limit because of the total spin conservation, there will be no transition from singlet to triplet states. As a result the scattering amplitude will be diagonal with respect to S_t and moreover will not depend upon μ_t. Thus in the total spin representation we have two amplitudes $\langle S_t|A|S_t\rangle$:

$$\langle 1|A|1\rangle = A^1, \quad \langle 0|A|0\rangle = A^0. \tag{1.20}$$

The relation between the amplitudes $\langle\mu_1'\mu_2'|A|\mu_1\mu_2\rangle$ and $\langle S_t|A|S_t\rangle$ is given by the standard transformation theory

$$\langle\mu_1'\mu_2'|A|\mu_1\mu_2\rangle = \sum_{S_t} C^{S_t\mu_t}_{\frac{1}{2}\mu_1'\frac{1}{2}\mu_2'} \langle S_t|A|S_t\rangle\, C^{S_t\mu_t}_{\frac{1}{2}\mu_1\frac{1}{2}\mu_2}. \tag{1.21}$$

$$\mu_t = \mu_1 + \mu_2 = \mu_1' + \mu_2'$$

Using the known values of the Clebsch-Gordan coefficients and adopting the notations (1.20), we get

$$\langle\mu\mu|A|\mu\mu\rangle = A^1 \tag{1.22}$$

$$\langle\mu,-\mu|A|\mu,-\mu\rangle = \frac{1}{2}(A^1 + A^0) \tag{1.23}$$

$$\langle-\mu,\mu|A|\mu,-\mu\rangle = \frac{1}{2}(A^1 - A^0). \tag{1.24}$$

Comparing (1.22) - (1.24) with (1.17) - (1.19) we can express a_1 and a_2 in terms of A^1 and A^0:

$$a_1 = \frac{1}{4}(3A^1 + A^0), \quad a_2 = \frac{1}{4}(A^1 - A^0). \tag{1.25}$$

Then (1.16) becomes

$$\hat{A} = \frac{1}{4}(3A^1 + A^0) + \frac{1}{4}(A^1 - A^0)(\sigma_1 \cdot \sigma_2). \qquad (1.26)$$

Finally, we will consider an alternative way of describing the spin state of a particle by indicating the spin projection on the unit vector along the direction of motion instead of the projection on some fixed direction.

The corresponding quantum number is called helicity. Before the collision the unit vector is $\mathbf{n}_0$, and the helicity λ_0 is defined by the equation

$$\hat{\mathbf{S}}\cdot\mathbf{n}_0|\lambda_0\rangle = \lambda_0|\lambda_0\rangle. \qquad (1.27)$$

After the collision the unit vector along the direction of motion will be $\mathbf{n}$, so

$$\hat{\mathbf{S}}\cdot\mathbf{n}|\lambda\rangle = \lambda|\lambda\rangle. \qquad (1.28)$$

Here λ_0 and λ are equal to $\pm\ 1/2$. But we will write + or - for any particular value of λ to avoid confusion with the particular values of μ which are also $\pm\ 1/2$.

The use of the helicity quantum number has some important advantages in the relativistic region which will be discussed below, and for that reason the helicity representation is adopted in elementary particle physics.

We will consider here only the case of a spin zero target. Let us establish the relation between the above amplitudes $\langle\mu'|\hat{A}|\mu\rangle$ and the helicity amplitudes $\langle\lambda|A|\lambda_0\rangle$. If we take $\mathbf{n}_0$ as the z-axis of the coordinate system then the state vector $|\lambda_0\rangle$ coincides with $|\mu\rangle$. However the final state vector refers to the direction $\mathbf{n}$, not $\mathbf{n}_0$. It follows that $\langle\lambda|$ does not coincide with any particular $\langle\mu'|$, but is rather a superposition of $\langle 1/2|$ and $\langle -1/2|$.

$$\langle +| = \cos\frac{\theta}{2}\langle 1/2| + \sin\frac{\theta}{2}\langle -1/2| \qquad (1.29)$$

$$\langle -| = \cos\frac{\theta}{2}\langle -1/2| - \sin\frac{\theta}{2}\langle 1/2| \qquad (1.30)$$

where θ is the angle between $\mathbf{n}$ and $\mathbf{n}_0$ in the scattering plane which will be taken as the (zx) plane. Using (1.29) and (1.30) we get the desired relations:

$$\langle +|A|+\rangle = \cos\frac{\theta}{2}\langle 1/2|A|1/2\rangle + \sin\frac{\theta}{2}\langle -1/2|A|1/2\rangle \qquad (1.31)$$

$$\langle -|A|-\rangle = \cos\frac{\theta}{2}\langle -1/2|A|-1/2\rangle - \sin\frac{\theta}{2}\langle 1/2|A|-1/2\rangle \qquad (1.32)$$

$$<+|A|-> = \cos\frac{\theta}{2}<1/2|A|-1/2> + \sin\frac{\theta}{2}<-1/2|A|-1/2> \quad (1.33)$$

$$<-|A|+> = \cos\frac{\theta}{2}<-1/2|A|1/2> - \sin\frac{\theta}{2}<1/2|A|1/2>. \quad (1.34)$$

Using (1.9) and taking into account that $\nu_x = 0$, $\nu_y = 1$ and $\nu_z = 0$, we get

$$<+|A|+> = <-|A|-> = \cos(\tfrac{\theta}{2})u + i\ \sin(\tfrac{\theta}{2})v \quad (1.35)$$

$$<+|A|-> = -<-|A|+> = \sin(\tfrac{\theta}{2})u\ - i\ \cos(\tfrac{\theta}{2})v. \quad (1.36)$$

Now, let us consider the problem of the partial wave expansion of the scattering amplitude. It is just this point where the use of helicity amplitudes has a great advantage over the amplitudes corresponding to the spin projection on some fixed direction. It turns out that the expansion of the helicity amplitude involves only the total angular momentum of the particle without decomposition into the orbital part and spin part, and has a very simple form

$$<\lambda|A|\lambda_o> = \sum_{j=\frac{1}{2}}^{\infty} (2j+1) <\lambda|A^j|\lambda_o> D^j_{\lambda\lambda_o}(\theta) \quad (1.37)$$

where

$$D^j_{\lambda\lambda_o}(\theta)$$

are the Wigner rotation functions.

2.2 THE DENSITY MATRIX FORMALISM

The scattering amplitudes considered in the previous section refer to an initial state with the definite spin projections on a given direction. But to describe the real experimental situation, an average of the cross section over an arbitrary initial spin state is needed.

The calculation of such an average value can be performed by means of the density matrix. Consider first a state $|a>$ which is a quantum superposition of $|1/2>$ and $|-1/2>$:

$$|a> = C_{\frac{1}{2}}|1/2> + C_{-\frac{1}{2}}|-1/2>. \quad (2.1)$$

The average value $\bar{L}$ of some operator $\hat{L}$ is equal to the diagonal matrix element

$$\bar{L} = <a|L|a>. \quad (2.2)$$

Using (2.1), we get

$$\bar{L} = \sum_{\mu,\mu'} C_\mu C^*_{\mu'} \langle\mu'|L|\mu\rangle \tag{2.3}$$

Now $C_\mu C^*_{\mu'}$ are in fact the matrix elements of an operator

$$\rho = |a\rangle\langle a| \tag{2.4}$$

which is the density operator. Then (2.3) can be represented in the form

$$\bar{L} = \sum_{\mu,\mu'} \langle\mu|\rho|\mu'\rangle\langle\mu'|L|\mu\rangle. \tag{2.5}$$

Taking into account that

$$\sum_{\mu'} \langle\mu|\rho|\mu'\rangle\langle\mu'|L|\mu\rangle = \langle\mu|\rho L|\mu\rangle,$$

we can reduce (2.5) to

$$\bar{L} = \sum_{\mu} \langle\mu|\rho L|\mu\rangle = \mathrm{Sp}(\rho\hat{L}). \tag{2.6}$$

The most general case of a state however is when there is no definite state vector at all, a particle beam being an example. In a real beam there is no definite spin state for all the particles together. Rather, there are random fluctuations in the spin projections over the beam due to thermal motion in the source, etc. This kind of state is called mixed in contrast to the superposition (2.1) which is called a pure state.

For any pure state the density matrix can be calculated by means of (2.4). For a mixed state there is no way of calculating the density matrix.

But we can define it by (2.6). This means we use (2.6) to express the density matrix in terms of the observed average values of some operators. The number of these operators should be equal to the number of parameters necessary to determine the density matrix. Then for all other quantities the average values can be calculated by means of (2.6). The previous consideration can be directly generalized to the case of a spin higher than 1/2 and also to the orbital angular momentum of an atom.

Having this in mind we will consider the basic properties of the density matrix for an arbitrary mixed state and establish the number of parameters needed to determine an N x N density matrix. We will follow the treatment given by Fano (1957). The first property of ρ is the normalization condition

$$\mathrm{Sp}\,\rho = 1. \tag{2.7}$$

It is an obvious consequence of (2.6) because the average value of 1 can be nothing else than 1.

The second property is a consequence of the requirement that the observed average value should be a real number. To ensure this, the density matrix should be Hermitian:

$$\rho = \rho^{+}. \tag{2.8}$$

Now let us count the number of independent parameters. The N x N matrix has N^2 elements, which are in general complex numbers, or $2N^2$ real parameters. The condition (2.8) reduces this number to N^2, and the condition (2.7) reduces it to $N^2 - 1$. So the N x N density matrix is determined by $N^2 - 1$ real parameters. For a pure state the number of parameters is equal to the number N of coefficients C. Each of the coefficients is a complex number, so we have 2N real parameters. The normalization condition reduces this number to 2N - 1. Another reduction comes from the possibility of dropping the common phase of all coefficients C because it does not enter in the observed values. Thus finally the number of independent real parameters in the density matrix for a pure state is 2N - 2. Now, it is easy to prove that

$$N^2 - 1 > 2N - 2$$

for any $N \geqslant 2$.

This difference between the number of independent parameters in mixed and pure states comes from the fact that in the case of a pure state there is a relation $\rho^2 = \rho$ which does not exist in the mixed state.

To get further insight into the structure of the density matrix, let us introduce N^2 operators which are represented by N x N matrices U_i having the following properties:

$$Sp(U_i U_k^{+}) = \delta_{ik}. \tag{2.9}$$

The matrices U_i need not be Hermitian, but if they are then $U_i^{+} = U_i$.

The density matrix can be represented as a linear superposition of U_i,

$$\rho = \sum_i \lambda_i U_i. \tag{2.10}$$

The meaning of λ_i can be established if we calculated $\overline{U_i^{+}}$ according to (2.2). Taking into account (2.9), we obtain

$$\lambda_i = \overline{U_i^{+}} .$$

It is convenient to choose one of the U_i to be the unit matrix multiplied by a normalization factor 1/N. (This can be done because the number of independent parameters should be $N^2 - 1$.) It follows then from (2.9) that

$$\mathrm{Sp}\, U_i = 0. \tag{2.11}$$

Two examples will be mentioned here which are considered in detail later. One is the set of Pauli matrices σ_x, σ_y, σ_z, which is used in the case of spin 1/2 particle (in this example U_i are in fact Hermitian). The second is the set of $(2j+1)^2$ multipole operators pertaining to the orientation of particles having angular momentum j. The multipole operators are conveniently chosen in tensorial form $T_{kq}(j)$ so that they transform under coordinate rotations like the spherical harmonics Y_{kq}, $q = -k \ldots k$ and $k = 0 \ldots 2j$. Each operator $T_{kq}(j)$ is represented by a $(2j+1)(2j+1)$ matrix. It is customary to choose the normalization condition in the form

$$\mathrm{Sp}[T^+_{kq}(j) T_{k'q'}(j)] = \delta_{kk'} \delta_{qq'} \tag{2.12}$$

and the phases according to

$$T^+_{kq} = (-1)^q T_{k,-q}. \tag{2.13}$$

However, in the literature one can meet with other conventions different from (2.12) and (2.13).

2.3 EXPRESSIONS FOR OBSERVABLE QUANTITIES

Now let us apply the general formalism outlined above to the collision process.

If there is no selection with respect to the spin projection, then the measured cross section will be a sum over the final spin projections and an average over the initial spin projections. First we will consider the case of a spin 0 target. Suppose for the time being that the initial spin projection is fixed, then the sum over the final spin projection will be equal to

$$Q = \sum_{\mu'} |\langle \mu' | A | \mu \rangle|^2. \tag{3.1}$$

We rewrite (3.1) in the form

$$Q = \sum_{\mu'} \langle \mu' | A | \mu \rangle^* \langle \mu' | A | \mu \rangle.$$

By definition of the Hermitian conjugation $\langle \mu' | A | \mu \rangle = \langle \mu | A^+ | \mu' \rangle$. So we have

$$Q = \sum_{\mu'} \langle \mu | A^+ | \mu' \rangle \langle \mu' | A | \mu \rangle = \langle \mu | A^+ A | \mu \rangle. \tag{3.2}$$

The expression $\hat{A}^+\hat{A}$ entering in (3.2) is the product of the operator (1.7) and its Hermitian conjugate. Because the matrices σ are Hermitian, we have

$$\hat{A}^+ = u^* + v^*(\boldsymbol{\sigma}\cdot\boldsymbol{\nu}),$$

so that

$$\hat{A}^+\hat{A} = [u^* + v^*(\boldsymbol{\sigma}\cdot\boldsymbol{\nu})][u + v(\boldsymbol{\sigma}\cdot\boldsymbol{\nu})].$$

If the initial state is an arbitrary superposition rather than a definite projection then the average over the initial projection will be $Q = \langle a|\hat{A}^+\hat{A}|a\rangle$. This expression has just the form (2.2) and therefore can be expressed in terms of the density matrix of the electronic beam like (2.6): $Q = \mathrm{Sp}(\rho\hat{A}^+\hat{A})$. Because of the invariance of Sp(...) against a cyclic permutation of operators, we rewrite the expression for Q as

$$Q = \mathrm{Sp}(\hat{A}\rho\hat{A}^+). \tag{3.3}$$

The same expression will hold in the case when the electronic beam is in a mixed state. (The electronic beam density matrix will be considered in detail in the next section.) Note that the expression

$$\rho' = \frac{1}{Q}\hat{A}\rho\hat{A}^+ \tag{3.4}$$

is in fact the final state density matrix. To verify this relation let us consider an initial spin state vector $|a\rangle$. The final state vector after collision will be $|b\rangle = N\hat{A}|a\rangle$ where N is a normalization factor. It follows that the final state density operator $\rho' = |b\rangle\langle b|$ is equal to $\rho' = |N|^2\hat{A}|a\rangle\langle a|\hat{A}^+ = |N|^2\hat{A}\rho\hat{A}^+$. To determine the factor $|N|^2$ we calculate the sum of diagonal matrix elements. Then

$$1 = |N|^2\mathrm{Sp}(\hat{A}\rho\hat{A}^+) = |N|^2 Q.$$

Using the matrix $\hat{\rho}'$ one can express an average value of any operator over the spin projection distribution after the collision

$$\bar{L} = \mathrm{Sp}(\hat{L}\rho'). \tag{3.5}$$

Let us now turn to the case of a spin 1/2 target. The same expressions (3.3) - (3.5) still hold provided that the density matrix for the composite system

$$\langle\mu_1'\mu_2'|\rho|\mu_1\mu_2\rangle$$

is used instead of $\langle\mu'|\rho|\mu\rangle$ and for the operator $\hat{A}$ the forms (1.13) or (1.16) are used instead of (1.7).

2.4 POLARIZATION OF A BEAM OF PARTICLES

The density matrix of an electronic beam can be constructed using the general expression (2.10). In our case N = 2; hence the matrices σ_x, σ_y, σ_z can be used in the role of U_i. Thus

$$\rho = \frac{1}{2}(1 + P_x\sigma_x + P_y\sigma_y + P_z\sigma_z). \tag{4.1}$$

According to (2.11)

$$P_{x,y,z} = \overline{\sigma_{x,y,z}}. \tag{4.2}$$

The numbers P_x, P_y, P_z are components of an axial vector **P** which characterizes the polarization of the beam. The number of components is just equal to the number of independent parameters needed to determine the most general 2 x 2 density matrix.

In the limiting case of a pure state

$$P^2 = 1 \tag{4.3}$$

[the relation (4.3) is a direct consequence of the general relation $\rho^2 = \rho$ which is in fact a definition of a pure state]. Because of the relation (4.3) the number of independent parameters drops from 3 to 2 just as it should be for a pure state. The beam is then completely polarized. When $P^2 < 1$ then the beam is partly polarized. In the limiting case P = 0 the beam is called unpolarized. One can say that the beam is completely chaotic (with relation to spin projection).

Now, let us consider an assembly of particles having angular momentum j. The density matrix can be represented in the form

$$\rho = \sum_{k=0}^{2j} \sum_{q=-k}^{k} P_q^k T_{kq}(j) \tag{4.4}$$

where $T_{kq}(j)$ are the tensor operators introduced at the end of Section 2.2.

If we adopt the normalization condition (2.12) then the matrix elements of T_{kq} are expressed in terms of the Clebsch-Gordan coefficients by the following relation

$$\langle\mu|T_{kq}(j)|\mu'\rangle = \sqrt{\frac{2k+1}{2j+1}}\, C^{j\mu}_{j\mu' kq}. \tag{4.5}$$

The numbers P_q^k are called polarization momenta. As an example, let us consider a beam of spin 1 particles.

There should be according to the general rule 8 independent parameters in the 3 x 3 density matrix for a mixed state. The polarization momenta P^1_{-1}, P^1_0, P^1_1 and P^2_{-2}, P^2_{-1}, P^2_0, P^2_1, P^2_2 form altogether just 8 parameters. The first three, $P^1_{-1,0,1}$, are the components of the polarization vector. The set of 5 momenta, $P^2_{-2,-1,0,1,2}$, characterize more complicated tensorial polarization. For instance, if the states with the angular momentum projection $j_z = \pm 1$ are equally populated but differ in population from $j_z = 0$, then there is no vector polarization. The beam is aligned and its alignment is characterized by the polarization momenta P^k_q.

Finally, we consider the composite assembly of a spin 1/2 particle beam and a spin 1/2 target. Usually the two parts - the beam and the target - are statistically independent (no correlations). Then the density matrix can be factorized

$$\langle\mu_1'\mu_2'|\rho|\mu_1\mu_2\rangle = \langle\mu_1'|\rho_1|\mu_1\rangle\,\langle\mu_2'|\rho_2|\mu_2\rangle \tag{4.5}$$

or in the matrix notation

$$\rho = \rho_1 \times \rho_2 \tag{4.6}$$

(the symbol × denotes a direct product). Using the explicit expression (4.1), we get

$$\rho = \frac{1}{4}(1 + \mathbf{P}_1\cdot\boldsymbol{\sigma}_1) \times (1 + \mathbf{P}_2\cdot\boldsymbol{\sigma}_2). \tag{4.7}$$

2.5 SCATTERING OF A PARTLY-POLARIZED BEAM

We are now able to study the scattering of a partly-polarized electronic beam on spin 0 and spin 1/2 targets. Let us begin with the spin 0 case. Assuming that the electronic beam before the scattering has a certain polarization using (3.3), (4.1) and (1.7) we have for the averaged cross section Q

$$Q = Q_0 + Q_1\mathbf{P}\cdot\boldsymbol{\nu} \tag{5.1}$$

where

$$Q_0 = |u|^2 + |v|^2 \tag{5.2}$$

$$Q_1 = 2|uv|\cos\gamma \tag{5.3}$$

$\gamma = \beta - \alpha$ [see (1.10)].

It appears that for a fixed scattering angle θ the cross section depends also upon the angle ϕ between the normal to the scattering plane and the polarization vector. At $\phi = 0$, $Q = Q_0 + Q_1P$, and at $\phi = \pi$, $Q = Q_0 - Q_1P$. So there is an asymmetry which is proportional to the value of the polarization vector P and therefore can be used to detect the presence of certain polarization in the beam. To deduce the numerical value of the polarization from the measured asymmetry one should calibrate the detector. The principles of the polarization detector construction and calibration can be found in the book "Polarized Electrons" by Kessler (1976). The polarization vector after the scattering is given by the expression

$$\mathbf{P}' = \mathrm{Sp}(\boldsymbol{\sigma}\rho') \tag{5.5}$$

which follows from (3.5). Inserting (1.7) and (4.1), we get

$$\mathbf{P}' = \frac{1}{Q}\{c_1\boldsymbol{\nu} + c_2\mathbf{P} + c_3[\boldsymbol{\nu},\mathbf{P}] + c_4\boldsymbol{\nu}(\boldsymbol{\nu}\cdot\mathbf{P})\} \tag{5.6}$$

where

$$\begin{aligned} c_1 &= 2|uv|\cos\gamma \\ c_2 &= |u|^2 - |v|^2 \\ c_3 &= 2|uv|\sin\gamma \\ c_4 &= 2|v|^2. \end{aligned} \tag{5.7}$$

Several facts can be established by inspection of (5.6) and (5.7). First, there is a nonzero polarization after the scattering even in the absence of any initial polarization. Putting P = 0, we have

$$\mathbf{P}' = \frac{2|u||v|\cos\gamma}{|u|^2 + |v|^2}\boldsymbol{\nu}. \tag{5.8}$$

It is seen that the polarization vector of an initially unpolarized beam is along $\boldsymbol{\nu}$. It acquires the maximum absolute value of unity if $|u| = |v|$ and $\cos\gamma = \pm 1$.

It follows from (1.7) that in the case $\cos\gamma = 1$,

$$\hat{A} = (1 + \boldsymbol{\sigma}\cdot\boldsymbol{\nu})u$$

and thus the amplitude corresponding to the spin projection $-1/2$ on the direction $\boldsymbol{\nu}$ is zero. In the case $\cos\gamma = -1$,

$$\hat{A} = (1 - \boldsymbol{\sigma}\cdot\boldsymbol{\nu})u$$

and thus the amplitude, corresponding to the spin projection $+1/2$ on the direction $\boldsymbol{\nu}$, is zero. Another interesting fact is the coefficients c_1 and Q are equal, so there is a direct relation between the polarization P' and asymmetry in the scattering of a polarized beam.

If $P \neq 0$, then it follows from (5.6) that in addition to the components of P' along P and $\boldsymbol{\nu}$, there will be also a component along $[P, \boldsymbol{\nu}]$. Now let us discuss the expressions (5.1) - (5.3) and (5.6) - (5.7) from the point of view of the information which can be obtained in various experiments: a) the parameter Q_o can be determined by the cross section measurement of an unpolarized beam; b) the ratio Q_1/Q_o can be determined either by measuring the asymmetry in the scattering of a polarized beam or by measuring the value of the polarization $\mathbf{P}'$ after the scattering of an unpolarized beam; and c) the ratios c_2/Q, c_3/Q or c_4/Q can be determined by measuring the correspondent components of the polarization vector $\mathbf{P}'$ along $\mathbf{P}$, $[\mathbf{P},\boldsymbol{\nu}]$ or $\boldsymbol{\nu}$ after the scattering of a polarized beam.

Next we consider the scattering of electrons by a spin 1/2 target in the nonrelativistic limit. Again we use the expression

$$Q = Sp(\hat{A}\rho\hat{A}^{+})$$

and this time insert (1.26) for $\hat{A}$ and (4.7) for $\hat{\rho}$. Then

$$Q = Q_o + Q_1\mathbf{P}_1\cdot\mathbf{P}_2 \tag{5.9}$$

where

$$Q_o = \frac{3}{4}|A^1|^2 + \frac{1}{4}|A^0|^2 \tag{5.10}$$

$$Q_1 = \frac{1}{4}(|A^0|^2 - |A^1|^2)\cos\phi \tag{5.11}$$

where ϕ is the difference between the phases of A^0 and A^1 (A^0 and A^1 are in general complex functions). The electronic polarization after the scattering is equal to

$$\mathbf{P}_1' = \frac{1}{Q}\, Sp(\sigma_1\hat{A}\rho\hat{A}^{+}). \tag{5.12}$$

The calculation gives

$$Q\mathbf{P}_1' = c_1\mathbf{P}_1 + c_2\mathbf{P}_2 + c_3[\mathbf{P}_1,\mathbf{P}_2] \tag{5.13}$$

where

$$c_1 = \frac{1}{2}(|A^1|^2 + |A^1||A^0|\cos\phi) \tag{5.14}$$

$$c_2 = \frac{1}{2}(|A^1|^2 - |A^1||A^0|\cos\phi) \tag{5.15}$$

$$c_3 = \frac{1}{4}(|A^0|^2 - |A^1|^2)\sin\phi. \tag{5.16}$$

It follows that: a) Q_o can be determined by measuring the scattering cross section of an unpolarized beam; b) Q_1 can be determined by measuring the cross section of a polarized beam on a polarized target; and c) the ratios c_1/Q, c_2/Q and c_3/Q can be determined by measuring the corresponding components of the electronic polarization along $\mathbf{P}_1$, $\mathbf{P}_2$ and $[\mathbf{P}_1,\mathbf{P}_2]$.

3
The Simplest Two-Channel System

3.1 FORMULATION OF THE PROBLEM: BOUNDARY CONDITIONS: S-MATRIX

The main difference between the scattering by a composite system such as an atom or molecule and by a central field is the possibility of the excitation of the target. Because of this the theory becomes much more complicated even for such a simple system as the hydrogen atom. It seems instructive in many respects to begin with a study of a simplified model of the target which can exist in only two states: the normal or ground state and just one excited state. Correspondingly we will speak about the elastic and inelastic scattering channels. Moreover, we will not take spin into account.

The short range interaction of the scattered particle with the target will be replaced by a boundary condition on a certain sphere.

The method of boundary conditions for a many-channel system in various modifications was developed by many authors. We will consider here the simplest version to make the most important features of many-channel systems as transparent as possible. The important advantage of the model adopted here is the possibility of taking into account the effect of the long-range interaction in the most direct and simple way.

There are four possible processes in the system under consideration: elastic scattering on the target in its normal state, excitation of the target, elastic scattering on an excited target and deexcitation of the target. Correspondingly we introduce the wave functions $F_{\alpha\beta}$ where α, β can have two possible values: 0 for the ground state and 1 for the excited state. β stands for the initial state, α for the final one. Let us denote by k_0 and k_1 the two values of the momentum of the particle involved in the collision. They are related to each other by an energy conservation condition

$$\frac{1}{2}k_0^2 - \frac{1}{2}k_1^2 = w \tag{1.1}$$

where w is the excitation energy. Outside the boundary sphere, $r > a$, the functions $F_{\alpha\beta}$ satisfy the equations

$$\begin{aligned} \Delta F_{0\beta} + k_0^2 F_{0\beta} &= 0 \\ \Delta F_{1\beta} + k_1^2 F_{1\beta} &= 0. \end{aligned} \tag{1.2}$$

If the target is initially in its normal state, then the asymptotic expressions for F_{00} and F_{01} are

$$\begin{aligned} F_{00} &\sim e^{i\mathbf{k}_0\cdot\mathbf{r}} + A_{00}\frac{e^{ik_0r}}{r} \\ F_{10} &\sim A_{10}\frac{e^{ik_1r}}{r}. \end{aligned} \tag{1.3}$$

If the target is initially in its excited state, then we have to consider the functions having the asymptotic expressions

$$\begin{aligned} F_{11} &\sim e^{i\mathbf{k}_1\cdot\mathbf{r}} + A_{11}\frac{e^{ik_1r}}{r} \\ F_{01} &\sim A_{01}\frac{e^{ik_0r}}{r}. \end{aligned} \tag{1.4}$$

By calculating the probability currents in the same fashion as in Section 1.1, we can express the cross sections in terms of the amplitudes

$$\frac{d\sigma_{\alpha\beta}}{d\Omega} = \frac{k_\alpha}{k_\beta}|A_{\alpha\beta}|^2. \tag{1.5}$$

The amplitudes $A_{\alpha\beta}$ satisfy two relations which are a direct generalization of (1.7) and (1.8) in Chapter 1. The first is

$$A_{10}(\mathbf{n},\mathbf{n}_0) = A_{01}(-\mathbf{n}_0,-\mathbf{n}). \tag{1.6}$$

By combining (1.5) and (1.6) we get the detailed balance relation

$$k_0^2\frac{d\sigma_{10}}{d\Omega} = k_1^2\frac{d\sigma_{01}}{d\Omega}. \tag{1.7}$$

The second is the conservation of the norm of the wave functions which has the meaning of the conservation of the total number of particles during the collisions

$$\int[F_{00}^*(\mathbf{k}_0,\mathbf{r})F_{00}(\mathbf{k}_0',\mathbf{r}) + F_{10}^*(\mathbf{k}_1,\mathbf{r})F_{10}(\mathbf{k}_1',\mathbf{r}) - e^{-i\mathbf{k}_0\cdot\mathbf{r}}e^{i\mathbf{k}_0'\cdot\mathbf{r}}]d\mathbf{r} = 0.$$

It follows that

$$\frac{2\pi}{k_0}\,\mathrm{Im}\,A_{00}(\mathbf{n}_o,\mathbf{n}_o) = \sum_\alpha \sigma_{\alpha 0}. \tag{1.8}$$

Let us perform the partial waves expansion:

$$F_{\alpha\beta} = \sum_\ell i^\ell \sqrt{4\pi(2\ell+1)}\,\frac{f_{\alpha\beta,\ell}}{r}\,Y_{\ell 0}(\theta).$$

We will suppose that the interaction does not mix the states with different angular momenta. Then each partial wave can be considered independently. The partial waves satisfy the equations

$$\left(\frac{d^2}{dr^2} + k^2 - \frac{\ell(\ell+1)}{r^2}\right)f_{\alpha\beta,\ell} = 0,\ (r > a).$$

The asymptotic expressions, corresponding to the target initially in the normal state, are

$$f_{00,\ell} \sim 1/k_0 \sin(k_0 r - \ell\pi/2) + A^\ell_{00}\,\mathrm{expi}(k_0 r - \ell\pi/2)$$

$$f_{10,\ell} \sim A^\ell_{10}\,\mathrm{expi}(kr - \ell\pi/2).$$

The amplitudes $A\alpha 0(\mathbf{n},\mathbf{n}_o)$ can be expressed in terms of the partial wave amplitudes A^ℓ_{00}, A^ℓ_{10}

$$A_{\alpha 0}(\mathbf{n},\mathbf{n}_o) = \sum_\ell (2\ell+1)A^\ell_{\alpha 0}P_\ell(\cos\theta). \tag{1.9}$$

For the cross sections we get

$$\sigma_{\alpha 0} = 4\pi\,\frac{k_\alpha}{k_0}\sum_\ell (2\ell+1)|A^\ell_{\alpha 0}|^2. \tag{1.10}$$

Now let us introduce the scattering matrix

$$S^\ell_{\alpha\beta} = \delta_{\alpha\beta} + 2i\sqrt{k_\alpha}A^\ell_{\alpha\beta}\sqrt{k_\beta} \tag{1.11}$$

and express the cross sections in terms of the scattering matrix elements

$$\begin{aligned}\sigma_{00} &= \frac{\pi}{k_0^2}\sum_\ell (2\ell+1)|S^\ell_{00} - 1|^2\\ \sigma_{10} &= \frac{\pi}{k_0^2}\sum_\ell (2\ell+1)|S^\ell_{10}|^2 .\end{aligned} \tag{1.12}$$

To take into account the conservation of the norm of the wave function, we insert (1.9) into (1.8) and get

$$A^\ell_{00} - (A^\ell_{00})^* = 2i[k_0|A^\ell_{00}|^2 + k_1|A^\ell_{10}|^2]. \tag{1.13}$$

Using (1.11) we get the corresponding relation for the scattering matrix elements

$$|S^\ell_{00}|^2 + |S^\ell_{10}|^2 = 1, \tag{1.14}$$

so the expression for σ_{10} acquires the form

$$\sigma_{10} = \frac{\pi}{k_0^2} \sum_\ell (2\ell + 1)(1 - |S_{00}^\ell|^2) \tag{1.15}$$

and we see that both elastic and inelastic scattering cross sections are expressed in terms of the same matrix element S_{00}^ℓ. Consider also the total cross section which is the sum of the elastic and inelastic cross sections

$$\sigma_{tot.} = \sigma_{00} + \sigma_{10}.$$

Its expression in terms of the scattering matrix is

$$\sigma_{tot.} = \frac{\pi}{k_0^2} \sum_\ell (2\ell + 1)(1 - \mathrm{Re}\, S_{00}^\ell). \tag{1.16}$$

Note that in contrast to σ_{00} and σ_{10} the total cross section is linear with respect to S_{00}^ℓ. For this reason the average over a small energy interval which sometimes is needed can be directly related to the average of S_{00}^ℓ for σ_{tot}, but cannot be directly related for σ_{00} and σ_{10}.

Let us consider in detail the case $\ell = 0$. For brevity the index ℓ will be dropped. It is advantageous to use matrix notation by introducing the matrices

$$f = \begin{pmatrix} f_{00} & f_{01} \\ f_{10} & f_{11} \end{pmatrix}, \quad k = \begin{pmatrix} k_0 & 0 \\ 0 & k_1 \end{pmatrix}. \tag{1.17}$$

The equations for the functions $f_{\alpha\beta}$ can be combined in one matrix equation

$$\frac{d^2}{dr^2} f + k^2 f = 0, \tag{1.18}$$

and so can the asymptotic expression

$$f = k^{-1} \sin kr + e^{ikr} A \tag{1.19}$$

where A is the matrix

$$A = \begin{pmatrix} A_{00} & A_{01} \\ A_{10} & A_{11} \end{pmatrix}.$$

Similar to the treatment in Section 1.2 of Chapter 1 we introduce the function R which has an asymptotic expression in the form

$$R = (2ik)^{-1}\{e^{ikr} I^* - e^{-ikr} I\}. \tag{1.20}$$

Here I is a direct generalization of the Jost function and will be called the Jost matrix. Then putting

$$f = RI^{-1},$$

inserting (1.20) and comparing with (1.19), we get the expression for A in terms of I:

$$A = (2ik)^{-1}(I^{*}I^{-1} - 1). \tag{1.21}$$

Now let us formulate the boundary conditions. They are imposed on the function R:

$$\left(\frac{dR}{dr}\right)_{r=a} = MR. \tag{1.22}$$

The matrix M represents the interaction of a particle with the target. It is real. To ensure the relation (1.6) it should also be symmetrical: $M_{01} = M_{10}$. Inserting (1.20) into (1.22) we have

$$I^{*}I^{-1} = e^{-ika}(M - ik)^{-1}(M + ik)e^{-ika}.$$

Supposing that $k_0 a << 1$, we will replace exp ika by 1 and write

$$I^{*}I^{-1} = (M - ik)^{-1}(M + ik).$$

Using (1.21) we get finally the expression for A in terms of M

$$A = \frac{1}{M - ik}. \tag{1.23}$$

It has just the form of relation (2.23) of Chapter 1 for elastic scattering. The difference is that M and A are matrices.

In our simple model the matrix elements of M are numbers. For any realistic multichannel system the amplitude matrix A can also be expressed in terms of a certain real symmetrical matrix M just in the form (1.23). However the elements of M will be functions of energy rather than numbers.

It follows from (1.23) that

$$A - A^{+} = 2iAkA^{+}. \tag{1.24}$$

Relation (1.13) is a consequence of (1.24).

The definition of the scattering matrix (1.11) can be rewritten in the matrix form

$$S = 1 + 2ik^{\frac{1}{2}}Ak^{\frac{1}{2}}. \tag{1.25}$$

It is easy to verify that (1.24) is equivalent to the condition of unitarity of the scattering matrix

$$SS^{+} = 1. \tag{1.26}$$

Taking into account (1.21), we can express the S-matrix in terms of I:

$$S = k^{-\frac{1}{2}} I^* I^{-1} k^{\frac{1}{2}}$$

and with the help of (1.23) the S-matrix is expressed in terms of M

$$S = k^{-\frac{1}{2}} \frac{M + ik}{M - ik} k^{\frac{1}{2}}. \tag{1.27}$$

The unitarity of S follows from (1.27) in a direct way. In the literature one often meets the closely related T-matrix.

$$T = k^{\frac{1}{2}} A k^{\frac{1}{2}} = k^{\frac{1}{2}} \frac{1}{M - ik} k^{\frac{1}{2}} \tag{1.28}$$

and K-matrix

$$K = k^{\frac{1}{2}} \frac{1}{M} k^{\frac{1}{2}}. \tag{1.29}$$

Now we turn to the partial waves $\ell \neq 0$. The matrix equation for f is

$$\frac{d^2}{dr^2} f + [k^2 - \frac{\ell(\ell+1)}{r^2}] f = 0. \tag{1.30}$$

The asymptotic form is

$$f \sim k^{-1} \sin(kr - \frac{\ell\pi}{2}) + e^{i(kr - \frac{\ell\pi}{2})} A. \tag{1.31}$$

The function R has by definition the asymptotic form

$$R \sim (2ik)^{-1} [e^{i(kr - \frac{\ell\pi}{2})} I^* - e^{-i(kr - \frac{\ell\pi}{2})} I].$$

At any finite r, R is expressed in terms of spherical Bessel functions

$$R = \frac{r}{2i} \{ j_\ell i (I^* + I) + (-1)^\ell j_{-\ell-1} (I^* - I) \}. \tag{1.32}$$

At small r, the function $j_{-\ell-1}$ behaves like $r^{-\ell}$. For that reason we formulate the boundary condition for Rr^ℓ rather than for R. Supposing that $k_0 a \ll 1$, we put

$$\left(\frac{(2\ell-1)!!}{a^\ell}\right)^2 \frac{d}{dr}(Rr^\ell)_{r=a} = MRa^\ell.$$

Inserting here (1.32), we get

$$I^* I^{-1} = (Mk^{-\ell-1} - ik^\ell)^{-1} (Mk^{-\ell-1} + ik^\ell). \tag{1.33}$$

For the amplitude A we obtain the expression

$$A = k^\ell (M - ik^{2\ell+1})^{-1} k^\ell \tag{1.34}$$

and for the S-matrix we obtain

$$S = k^{-\ell-\frac{1}{2}} \frac{M + ik^{2\ell+1}}{M - ik^{2\ell+1}} k^{\ell+\frac{1}{2}}. \tag{1.35}$$

3.2 ELASTIC SCATTERING BELOW THE THRESHOLD. RESONANCES

It was assumed in the previous section that $k_0^2 \geqslant 2w$ so that both channels are open. Let us now consider the case where $k_0^2 \leqslant 2w$ so that the only process which is possible is elastic scattering. In this case k_1 will be purely imaginary and we put

$$\kappa = \frac{1}{i} k_1 = \sqrt{2w - k_0^2} .$$

If $k_1^2 < 0$ then the functions R_{11} and R_{10} contain exponentially growing terms as $r \to \infty$. The only way to be rid of them is to put $R_{11} = 0$, $R_{10} = 0$.

Then f_{11} and f_{10} also will be $= 0$. It follows that the only amplitude which exists is A_{00}. But the boundary conditions for R_{00} and R_{01} will still involve all 4 elements of the matrix M and so will the amplitude A_{00}. In this sense it can be said that the closed channels affect the open elastic scattering channel.

The S-matrix will reduce to a single number S_{00}^{ℓ} and the unitary condition requires that

$$|S_{00}^{\ell}|^2 = 1. \tag{2.1}$$

It follows that S_{00}^{ℓ} can be represented in the form

$$S_{00}^{\ell} = e^{2i\delta_\ell}, \tag{2.2}$$

similar to potential scattering.

Let us consider in detail the partial wave $\ell = 0$. We have

$$e^{2i\delta_0} = \frac{(M_{00} + ik_0)(M_{11} + \kappa) - M_{01}^2}{(M_{00} - ik_0)(M_{11} + \kappa) - M_{01}^2} .$$

It follows that

$$k_0 \cot\delta_0 = M_{00} - \frac{M_{01}^2}{M_{11} + \kappa} . \tag{2.3}$$

The first term on the right side of (2.3) corresponds to potential scattering. The influence of the closed channel is represented by the second term. The strength of this influence is determined by M_{01}. The most interesting consequence of this influence is the possibility of resonances. Similar to the treatment of resonances in the potential scattering, we split the phase δ_0 into two parts

$$\delta_0 = \delta_0' + \delta_0'' \tag{2.4}$$

and define δ_0' by the expression

$$e^{2i\delta_0'} = \frac{M_{00} + ik_0}{M_{00} - ik_0} . \tag{2.5}$$

The phase δ_0' describes the scattering in the case $M_{01} = 0$, that is in the absence of coupling between channels. For the rest we get

$$e^{2i\delta_0''} = [1 - \frac{M_{01}^2}{(M_{00} + ik_0)(M_{11} + \kappa)}][1 - \frac{M_{01}^2}{(M_{00} - ik_0)(M_{11} + \kappa)}]^{-1} . \tag{2.6}$$

It follows from (2.6) that

$$k_0 M_{01}^2 \cot\delta_0'' = (M_{11} + \kappa)(M_{00}^2 + k_0^2) - 2M_{01}^2 M_{00} \tag{2.7}$$

Let us consider the case of a weak coupling of the channels (M_{01} is small). Then

$$\cot\delta_0'' \quad \frac{M_{00}^2 + k_0^2}{M_{01}^2 k_0}(M_{11} + \kappa) . \tag{2.8}$$

It is seen that $\cot\delta_0''$ becomes zero if $M_{11} + \kappa = 0$.

To understand the meaning of this condition, let us consider the scattering on an excited target in the absence of the coupling ($M_{01} = 0$). Then the amplitude A_{11} will be

$$A_{11} = \frac{1}{M_{11} - ik_1} .$$

At imaginary $k_1 = i\kappa$, we have

$$A_{11} = \frac{1}{M_{11} + \kappa} .$$

The condition $M_{11} + \kappa = 0$ gives a pole in A_{11} which corresponds to a bound state in the field of an excited target.

In the presence of the coupling this state will not be a true bound state but rather a quasistationary one because the target will return to its normal state and the temporarily bound particle will decay. Correspondingly, the analytical continuation of the amplitude A_{11} below the threshold will have complex poles instead of pure imaginary poles.

The position of the complex poles of A_{11} is given by the relation

$$(M_{11} - ik_1)(M_{00} - ik_0) - M_{01}^2 = 0.$$

To get an approximate expression for the width Γ in the case of weak coupling we consider the expression (2.8) near the energy E_0 at which $\cot\delta_0''$ is zero. It has the form

$$\cot\delta_0'' = \frac{M_{00}^2 + k_0^2}{M_{01}^2 |M_{11}| k_0}(E_0 - E). \tag{2.9}$$

Comparing this with the standard expression

$$\cot\delta'' = 2(E_0 - E)/\Gamma,$$

we see that

$$\Gamma = 2|M_{11}| k_0 \frac{M_{01}^2}{M_{00}^2 + k_0^2}. \tag{2.10}$$

It follows that in the case of weak coupling the resonance will be narrow.

3.3 CUSPS

There is an interesting feature of the elastic scattering cross section at the inelastic threshold: it has an infinite derivative on the energy scale.

The four possible shapes of $d\sigma_{00}/d\Omega$ near the threshold are shown in Figure 3.1, both on the energy scale and momentum scale. This feature is entirely due to the $\ell = 0$ partial amplitudes A_{00}^0 and A_{10}^0. Let us write down the amplitude A_{00}^0 below the threshold and the amplitudes A_{00}^0 and A_{10}^0 above the threshold. Below the threshold:

$$A_{00}^0 = \frac{M_{11} + \kappa}{(M_{00} - ik_0)(M_{11} + \kappa) - M_{01}^2}. \tag{3.1}$$

Above the threshold:

$$A_{00}^0 = \frac{M_{11} - ik_1}{(M_{00} - ik_0)(M_{11} - ik_1) - M_{01}^2},$$

$$A_{10}^0 = \frac{M_{10}}{(M_{00} - ik_0)(M_{11} - ik_1) - M_{01}^2}. \tag{3.2}$$

Now we expand A_{00}^0 and A_{10}^0 in powers of k_1 taking into account the zero and the first powers only, and denote all the quantities at the threshold by the index t.

Then above the threshold

$$A_{00}^0 = (A_{00}^0)_t + ik_1 |A_{10}|_t^2 (e^{2i\delta_0})_t, \tag{3.3}$$

and below the threshold

$$A_{00}^0 = (A_{00}^0)_t - \kappa |A_{10}^0|_t^2 (e^{2i\delta_0})_t. \tag{3.4}$$

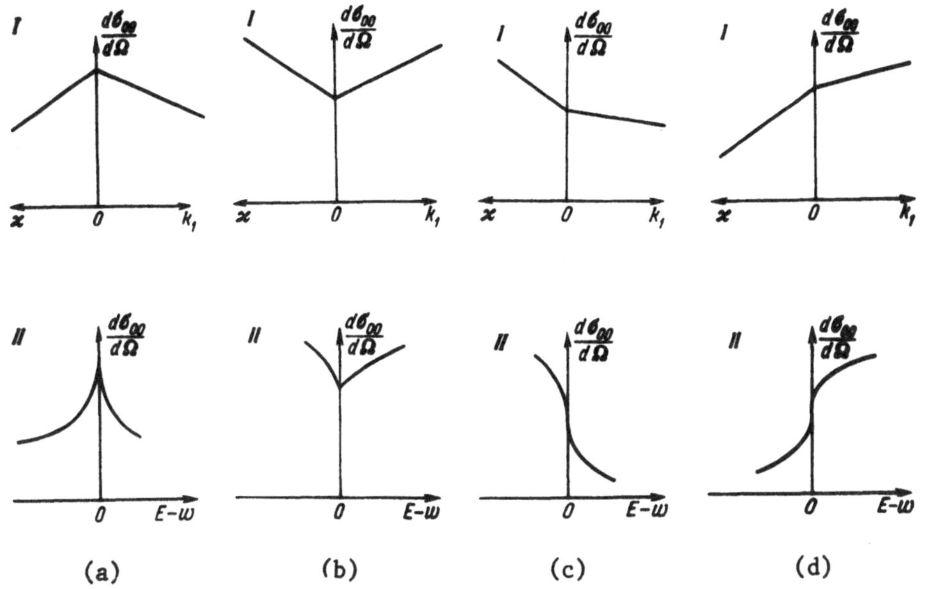

Fig. 3.1 Cusps and rounded steps on an energy scale (II) and corresponding salient features on a momentum scale (I).

(a) $\mathrm{Im}(A_{00}e^{-2i\delta_0}) < 0$, $\mathrm{Re}(A_{00}e^{-2i\delta_0}) > 0$;

(b) $\mathrm{Im}(A_{00}e^{-2i\delta_0}) > 0$, $\mathrm{Re}(A_{00}e^{-2i\delta_0}) < 0$;

$\mathrm{Im}(A_{00}e^{-2i\delta_0}) < 0$, $\mathrm{Re}(A_{00}e^{-2i\delta_0}) < 0$;

$\mathrm{Im}(A_{00}e^{-2i\delta_0}) > 0$, $\mathrm{Re}(A_{00}e^{-2i\delta_0}) > 0$.

For higher partial waves $\ell = 0$ there will be no terms linear in k_1, so we put

$$A_{00}^{\ell} = |A_{00}^{\ell}|_t. \tag{3.5}$$

Inserting into (1.9) all these expressions, we get, for above the threshold,

$$A_{00}(\theta) = [A_{00}(\theta)]_t + ik_1|A_{10}^{0}|_t^2(e^{2i\delta_0})_t, \tag{3.6}$$

and below the threshold,

$$A_{00}(\theta) = [A_{00}(\theta)]_t - \kappa|A_{10}^{0}|_t^2(e^{2i\delta_0})_t. \tag{3.7}$$

For the differential cross section

$$\frac{d\sigma_{00}}{d\Omega} = |A_{00}(\theta)|^2,$$

we have, for above the threshold

$$\frac{d\sigma_{00}}{d\Omega} = \left(\frac{d\sigma_{00}}{d\Omega}\right)_t + 2k_1|A^0_{10}|^2_t \mathrm{Im}(A_{00}(\theta)e^{2i\delta_0})_t, \tag{3.8}$$

and below the threshold

$$\frac{d\sigma_{00}}{d\Omega} = \left(\frac{d\sigma_{00}}{d\Omega}\right)_t - 2\kappa|A^0_{10}|^2_t \mathrm{Re}(A_{00}(\theta)e^{2i\delta_0})_t. \tag{3.9}$$

The shape of the curves shown in Figure 3.1 depends on the signs of $\mathrm{Re}(A_{00}(\theta)e^{2i\delta_0})_t$ and $\mathrm{Im}(A_{00}(\theta)e^{2i\delta_0})_t$ which is indicated in the caption.

It follows from the expressions (3.8) and (3.9) that the experimental study of the cusps can provide information about $|A^0_{10}|^2_t$ and the phase of $A_{00}(\theta)e^{2i\delta_0}$. Integrating (3.8) and (3.9) over the angles and taking into account that

$$\int A_{00}(\theta)d\Omega = A^0_{00},$$

we get, for above the threshold

$$\sigma_{00} = (\sigma_{00})_t - 8\pi\frac{k_1}{k_0}|A^0_{10}|^2_t(\sin^2\delta_0)_t, \tag{3.10}$$

and below the threshold

$$\sigma_{00} = (\sigma_{00})_t - 8\pi\frac{\kappa}{k_0}|A^0_{10}|^2_t(\cos\delta_0\sin\delta_0)_t. \tag{3.11}$$

It follows that for the total cross section there are only two possible shapes: 3.1a and 3.1b.

3.4 CROSS SECTIONS ABOVE THE THRESHOLD

We will consider the $\ell = 0$ partial wave cross sections for elastic scattering and excitation. First let us write down the values of $|A_{00}|^2$ and $|A_{10}|^2$. It follows from (3.2) that

$$|A_{00}|^2 = \frac{M^2_{11} + k^2_1}{(M_{00}M_{11} - M^2_{01} - k_0k_1)^2 + (k_0M_{11} + k_1M_{00})^2} \tag{4.1}$$

$$|A_{10}|^2 = \frac{M^2_{10}}{(M_{00}M_{11} - M^2_{01} - k_0k_1)^2 + (k_0M_{11} + k_1M_{00})^2}. \tag{4.2}$$

To get a further insight into these expressions we consider the eigenphases of the S-matrix. By a unitary transformation U one can transform the S-matrix to the diagonal form

$$\tilde{S} = USU^{-1}. \tag{4.3}$$

Because of the unitarity of the S-matrix the diagonal elements of $\tilde{S}$ can be expressed in the form $e^{2i\Delta_1}$ and $e^{2i\Delta_2}$ so that

$$\tilde{S} = \begin{pmatrix} e^{2i\Delta_1} & 0 \\ 0 & e^{2i\Delta_2} \end{pmatrix}. \tag{4.4}$$

Δ_1 and Δ_2 are called the eigenphases of the S-matrix.

It follows that the determinant of the $\tilde{S}$-matrix (which is the same as that of the S-matrix) is equal to $e^{2i(\Delta_1 + \Delta_2)}$. On the other hand, it follows from (1.26) that this determinant is equal to

$$||S|| = \frac{(M_{00} + ik_0)(M_{11} + ik_1) - M_{01}^2}{(M_{00} - ik_0)(M_{11} - ik_1) - M_{01}^2}. \tag{4.5}$$

Thus

$$\cot(\Delta_1 + \Delta_2) = \frac{M_{00}M_{11} - M_{01}^2 - k_0k_1}{k_1M_{00} + k_0M_{11}}. \tag{4.6}$$

Taking into account (4.6) we can rewrite (4.1) and (4.2) in the form

$$|A_{00}|^2 = \frac{M_{11}^2 + k_1^2}{(k_0M_{11} + k_1M_{00})^2} \sin^2(\Delta_1 + \Delta_2) \tag{4.7}$$

$$|A_{10}|^2 = \frac{M_{01}^2}{(k_0M_{11} + k_1M_{00})^2} \sin^2(\Delta_1 + \Delta_2). \tag{4.8}$$

It can be seen that both in (4.7) and (4.8) there is a common factor $\sin^2(\Delta_1 + \Delta_2)$. If $M_{00}M_{11} - M_{01}^2 > 0$, then at some energy $\cot(\Delta_1 + \Delta_2)$ can pass through zero from positive to negative values, just as in the case of a resonance in potential scattering. Then the factor $\sin^2(\Delta_1 + \Delta_2)$ will have a more or less sharp peak at the point where

$$\Delta_1 + \Delta_2 = \frac{\pi}{2}. \tag{4.9}$$

It follows that in this respect the sum of the S-matrix eigenphases plays the same role in the many-channel system as the phase δ does in potential scattering.

If other factors in (4.7) and (4.8) do not change significantly over the peak, then there would be a resonance in both channels. However, close inspection of these factors shows that this is not the case at least for the excitation cross section.

The expression (4.2) can be rewritten in the form

$$|A_{10}|^2 = \frac{M_{10}^2}{(M_{11}M_{00} - M_{01})^2 + k_0^2 M_{11}^2 + k_1^2 M_{00}^2 + k_0^2 k_1^2} .$$

It is easy to see that $|A_{00}|^2$ is monotonically increasing and acquires its maximum value at threshold where

$$|A_{10}|^2_{k_1=0} = \frac{M_{10}^2}{(M_{00}M_{11} - M_{01}^2)^2 + k_0^2 M_{11}^2} = \frac{M_{10}^2}{k_0^2 M_{11}^2} \sin^2(\Delta_1 + \Delta_2). \tag{4.10}$$

The derivative

$$\left(\frac{d|A_{10}|^2}{dk_1}\right)_{k_1=0} = -2k_0|A_{10}|^2. \tag{4.11}$$

It follows that the larger $|A_{10}|^2$ is, the more sharp will be the decrease. The favorable condition for $|A_{10}|^2_{k_1=0}$ to be large is again

$$(\Delta_1 + \Delta_2)_{k_1=0} = \frac{\pi}{2} .$$

Turning to the cross section we should not forget that there is a factor k_1/k_0 in front of $|A_{10}|^2$ in expression (1.5). So the excitation cross section will have a sharp peak near $k_1 = 0$. It should be noted that in the real multichannel systems where $M_{\alpha\beta}$ are functions of the energy rather than numbers, a resonance due to condition (4.9) can happen in the excitation channels too.

Finally we just mention the consequence of (1.34) that the partial amplitude A_{10}^{ℓ} in the vicinity of the threshold is proportional to k_1^{ℓ} so that the corresponding partial cross section will be proportional to $k_1^{2\ell+1}$.

The relation

$$\sigma_{10} \sim k_1^{2\ell + 1} \tag{4.12}$$

is called Wigner's threshold law. It is important to realize that the energy interval where (4.12) holds can be in fact very narrow and outside this interval the interrelation between the partial cross sections for various ℓ can be very different from (4.12).

3.5 EFFECT OF AN ATTRACTIVE COULOMB FIELD

The effect of an attractive Coulomb field is very important in collisions of an electron with a positive atomic ion. The obvious result is the occurrence of an infinite hydrogen-like series of the resonances in the elastic scattering cross section below the excitation thresholds. They correspond to the bound states of an electron in the field of an excited positive ion.

The inclusion of the effect of an attractive Coulomb field in our simple two-channel system is achieved by using instead of $\exp(\pm ikr)$ the Coulomb wave functions $\phi\pm$ having the asymptotic forms

$$\exp[\pm i(kr + \frac{Z}{k}\ln 2kr + \eta_0)].$$

We will follow the treatment given by Landau and Lifshitz (1974) and will consider only the case $\ell = 0$. Assuming that $ka \ll 1$ and $Z/k \gg 1$, it can be shown that instead of (1.26) we will have for the S-matrix the following expression

$$S = k^{-\frac{1}{2}}(M - \lambda^*)(M - \lambda)^{-1}k^{\frac{1}{2}} \tag{5.1}$$

where

$$\lambda = \left(\frac{1}{\phi^+}\frac{d\phi^+}{dr}\right)_{r=a}. \tag{5.2}$$

The element S_{00} relevant to elastic scattering is equal to

$$S_{00} = \frac{[M_{00} - \lambda^*(k_0)][M_{11} - \lambda(k_1)] - M_{10}^2}{[M_{00} - \lambda(k_0)][M_{11} - \lambda^*(k_1)] - M_{01}^2}. \tag{5.3}$$

Now let us perform an analytical continuation of this expression below the threshold and put $k_1 = i\kappa$. It follows from the known properties of the Coulomb wave function that for small and real k

$$\lambda(k) = i + \text{const}, \tag{5.4}$$

and for imaginary $k_1 = i\kappa$

$$\lambda(i\kappa) = -\cot\frac{\pi Z}{\kappa} + \text{const}. \tag{5.5}$$

The constant terms in (5.4) and (5.5) can be absorbed in M_{00} and M_{11}. Then below the threshold we have

$$e^{2i\delta_0} = \frac{(M_{00} + i)[M_{11} + \cot(\pi Z/\kappa)] - M_{10}^2}{(M_{00} - i)[M_{11} + \cot(\pi Z/\kappa)] - M_{10}^2}. \tag{5.6}$$

It can be seen that the phase δ_0 becomes $\pi/2$ under the condition

$$\cot(\pi Z/\kappa) = \frac{M_{01}^2}{M_{00}} - M_{11}. \tag{5.7}$$

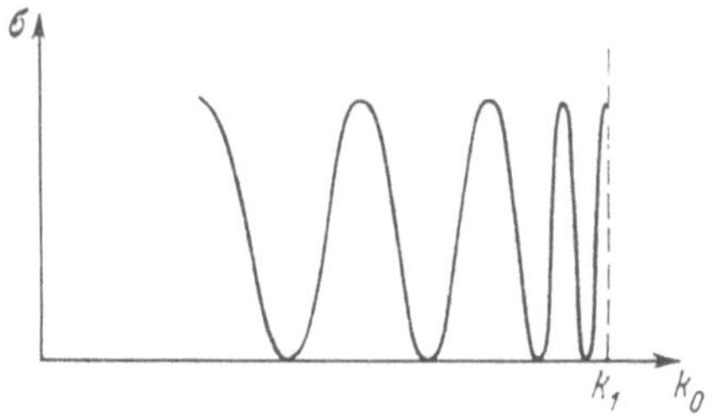

Fig. 3.2 Resonances below the excitation threshold in the presence of Coulomb field.

So we get a series of resonances as was expected. The resonances are equidistant on the scale $1/\kappa$ but on the κ scale they are condensing towards the threshold (Figure 3.2). Note that for a pure Coulomb potential $\cot(\pi Z/\kappa)$ would be infinite.

There is an alternative way of reproducing the effect of an attractive Coulomb field in which $\cot\delta_0$ is considered rather than $\exp 2i\delta_0$. It follows from (5.6) that

$$\cot\delta_0 = M_{00} - \frac{M_{10}^2}{M_{11} + \cot(\pi Z/\kappa)} . \tag{5.8}$$

This expression is similar to (2.3) with the exception that instead of $k\cot\delta_0$ we have here $\cot\delta_0$, and instead of κ we have $\cot(\pi Z/\kappa)$. The expression (5.8) can be interpreted as an analytical continuation of the M-matrix below the excitation threshold. In fact it is just the idea of the many-channel quantum defect theory developed by Seaton (1969). The only difference is that Seaton considered the K-matrix which is equal to M^{-1}.

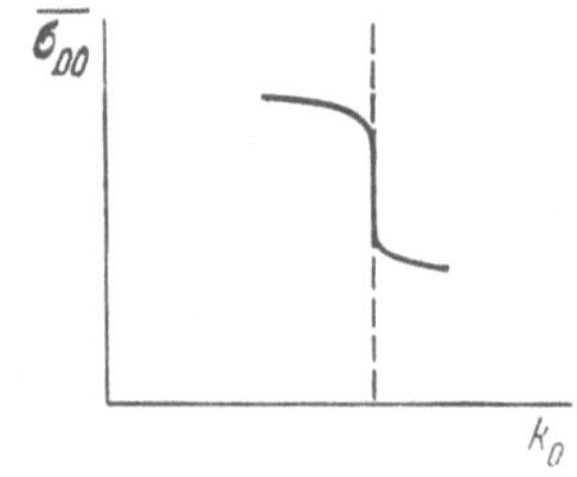

Fig. 3.3 Threshold step of the averaged elastic scattering cross section.

In the simplest case of a two-channel system, considered here, Seaton's relation can be written in the form

$$\tilde{K} = K_{00} - K_{01}\left(K_{11} + \tan\frac{\pi Z}{\kappa}\right)^{-1} K_{10} \tag{5.9}$$

where $\tilde{K}$ is the reactance matrix below the threshold (which is equal to $\tan\delta_0$). In the vicinity of the threshold the cross section averaged over a small energy interval is more relevant to the real experimental conditions (because of the very rapid oscillations).

It is interesting that the averaged elastic scattering cross section below the threshold (but very near to it) is equal to the sum of the elastic and inelastic scattering cross section, so that the elastic scattering cross section itself will have a step just at the threshold (Gailitis, 1963; Baz, 1959) (Figure 3.3).

4
Collisions of Electrons with Atoms and Ions: General Theory

4.1 WAVE FUNCTION OF THE SYSTEM: AMPLITUDES AND CROSS SECTIONS

The wave function of the system "electron + atom" depends upon electronic coordinates $\mathbf{r}_1 \ldots \mathbf{r}_N$ and spin projections $\mu_1 \ldots \mu_N$. It should be antisymmetric with respect to the permutation of any pair $\mathbf{r}_i\mu_i$ and $\mathbf{r}_k\mu_k$. Denoting by one quantity x_i all variables $\mathbf{r}_i\mu_i$ and by P_{ik} the permutation operator, we have

$$P_{ik}\Phi(\ldots,x_i\ldots,x_k\ldots) = -\Phi(\ldots,x_i\ldots,x_k\ldots). \tag{1.1}$$

The function Φ is an eigenfunction of the Hamiltonian operator H

$$H\Phi = E\Phi \tag{1.2}$$

where H in the nonrelativistic approximation has the form

$$H = -\frac{1}{2}\sum_i \Delta_i - \sum_i \frac{Z}{r_i} + \sum_{i<j} \frac{1}{|\mathbf{r}_i - \mathbf{r}_j|}. \tag{1.3}$$

The first term on the right side is the kinetic energy of all the electrons, the second term is the interaction energy of the electrons with the nuclei, and the third term is the interaction energy between the electrons.

The energy E is the sum of the incident electron energy $\frac{1}{2}k_0^2$ and that of the atom E_0:

$$E = E_0 + \frac{1}{2}k_0^2. \tag{1.4}$$

Besides equation (1.2), Φ should satisfy certain asymptotic conditions when any of the coordinates, say r_N, tend to infinity.

The asymptotic conditions will be formulated in terms of atomic wave functions $\phi_{nLMS\mu_s}$, the spin part of the scattered electron wave function u_μ and the coordinate part of the scattered electron wave function (which consists of a plane wave in the entrance channel and outgoing spherical waves in all open channels). The quantum number which specifies the atomic wave function has the following meaning: n is the number of the energy level, L and M are the angular momentum and its projection quantum numbers, S and μ_s are the atomic spin and its projection. For brevity we denote $\alpha = (n,L,M,S)$. The asymptotic condition in question has the form

$$\Phi \sim e^{i\mathbf{k}_0\cdot\mathbf{r}} u_{\mu^0}\phi_{\alpha^0\mu_s^0} + \sum_{\alpha\mu_s\mu} \frac{e^{ik_\alpha r}}{r} u_\mu \phi_{\alpha\mu_s} \langle\alpha\mu_s\mu|A|\alpha^0\mu_s^0\mu^0\rangle. \quad (1.5)$$

Here $\alpha^0\mu_s^0\mu^0$ denote the initial state. r is written instead of r_N. The summation over α includes the open channels only. The values k_0 and k_α are related by the energy conservation

$$\frac{1}{2}k_0^2 + E_0 = \frac{1}{2}k_\alpha^2 + E_\alpha, \quad (1.6)$$

where E_α and E_α are the atomic energy levels.

In the case of a collision with an atomic ion of charge Z_0, $\mathbf{k}_0\cdot\mathbf{r}$ and $k_\alpha r$ must be replaced by the expressions

$$\mathbf{k}_0\cdot\mathbf{r} - \frac{Z_0}{k_0}\ln(2k_0 r\sin^2\theta/2),$$

and

$$k_\alpha r + \frac{Z_0}{k_\alpha}\ln(2k_\alpha r).$$

The differential cross section for the transition $\alpha^0\mu^0\mu_s^0 \to \alpha\mu\mu_s$ is given by the expression

$$\frac{d\sigma_{\alpha\mu_s\mu\leftarrow\alpha^0\mu_s^0\mu^0}}{d\Omega} = \frac{k_\alpha}{k_0}\,|\langle\alpha\mu_s\mu|A|\alpha^0\mu_s^0\mu^0\rangle|^2. \quad (1.7)$$

The T-matrix is often used instead of the amplitude A. They are related in the following way

$$\langle\alpha\mu_s\mu|T|\alpha^0\mu_s^0\mu^0\rangle = \sqrt{k_\alpha k_0}\,\langle\alpha\mu_s\mu|A|\alpha^0\mu_s^0\mu^0\rangle.$$

The expression for the cross section in terms of T has the form

$$\frac{d\sigma_{\alpha\mu_s\mu\leftarrow\alpha^0\mu_s^0\mu^0}}{d\Omega} = \frac{1}{k_0^2}\,|\langle\alpha\mu_s\mu|T|\alpha^0\mu_s^0\mu^0\rangle|^2. \quad (1.8)$$

The amplitude $\langle \alpha\mu_s\mu|A|\alpha^0\mu_s^0\mu^0\rangle$ obeys an important relation which follows from the invariance against reflection in the scattering plane (Macek and Jaecks, 1971)

$$\langle LMS\mu_s\mu|A|L^0M^0S^0\mu_s^0\mu^0\rangle$$
$$= (-1)^{M-M^0+S-S^0+\mu_s-\mu^0+\mu-\mu^0}\langle L,-M,S,-\mu_s,-\mu|A|L^0,-M^0,S^0,-\mu_s^0,-\mu\rangle . \qquad (1.9)$$

It will be used in Section 4.7.

The use of projectile and target spin projections μ_s,μ is directly related to the experimental conditions because these quantities can be measured. But for the theoretical treatment it is advantageous to use the representation in which the total spin S_t at its projection μ_t are used because in the nonrelativistic approximation these quantities are conserved during the collision process. The amplitude in the total spin representation is diagonal in S_t and is independent of μ_t (as far as the nonrelativistic case is concerned). Taking this into account we will denote the amplitude by

$$\langle\alpha|A^{S_t}|\alpha^0\rangle .$$

The relation between the amplitudes in both representations is given by the standard transformation theory

$$\langle\alpha\mu_s\mu|A|\alpha^0\mu_s^0\mu^0\rangle = \sum_{S_t} C^{S_t\mu_t}_{S\mu_s\frac{1}{2}\mu} C^{S_t\mu_t}_{S^0\mu_s^0\frac{1}{2}\mu^0}\langle\alpha|A^{S_t}|\alpha^0\rangle . \qquad (1.10)$$

It follows from total spin conservation that

$$\mu_t = \mu_s + \mu = \mu_s^0 + \mu^0 .$$

In the case of a transition without change of the target spin ($S = S^0$) the sum in (1.10) contains two terms, corresponding to $S_t = S^0 - \frac{1}{2}$ and $S_t = S^0 + \frac{1}{2}$ except for the case $S^0 = 0$, when the sum reduces to one term only. For transitions involving a change of the target spin there is also one term in the sum (1.10). If $S = S^0 + 1$, then $S_t = S^0 + \frac{1}{2}$. If $S = S^0 - 1$, then $S_t = S^0 - \frac{1}{2}$.

So far we have discussed the scattering amplitude as a product of theoretical calculation. Now let us look on the amplitude from the standpoint of experiment. To what extent can we reconstruct the dependence of the amplitude upon the projection of atomic spin μ_s, the orbital angular momentum M, and the spin projection μ of the incident electron (provided that all other quantum numbers are fixed), using experimental data?

To reconstruct the dependence upon the spin projections μ_s and μ, spin-polarization measurements are needed like those discussed in Chapter 2 for elastic scattering. If there are no polarization measurements then the observed cross section will be a sum over the final spin projections μ_s, μ and an average over the initial ones μ_s, μ. In the case of unpolarized beams this cross section is equal to

$$\bar{Q} = \frac{1}{2(2S^0+1)} \sum_{\mu_s \mu_s^0 \mu \mu^0} |\langle \alpha \mu_s \mu | A | \alpha^0 \mu_s^0 \mu^0 \rangle|^2. \tag{1.11}$$

Now using (1.10), taking into account the properties of the Clebsch-Gordan coefficients, and summing over all μ_t, we get

$$\bar{Q} = \frac{1}{2(2S^0+1)} \sum_{S_t} (2S_t+1) |\langle \alpha | A^{S_t} | \alpha^0 \rangle|^2. \tag{1.12}$$

The reconstruction of the dependence upon μ at least for an optically-allowed transition could be done to a certain extent by observation of the light emitted by an excited atom. However, it is important to realize that the optical measurements should be accompanied by measurements of the scattered electron motion in order to fix the position of the scattering plane. In other words an electron-photon coincidence experiment is needed, otherwise the light observed in the experiment will implicitly correspond to the average over all orientations of the scattering plane. Because of this averaging, some information will be lost.

It follows that the study of correlation phenomena in electron-photon coincidence experiments forms an important part of atomic collision physics.

The theoretical and experimental aspects of this problem are discussed in a review article by Blum and Kleinpoppen (1979) in the book "Coherence and Correlation in Atomic Collision" edited by Kleinpoppen and Williams (1980) and in the book by Blum (1981).

Some of the basic ideas will be considered later in Section (4.7).

4.2 THE TOTAL SPIN REPRESENTATION. SEPARATION OF THE SPIN VARIABLES

Let us consider the wave function of an N-electron system Φ_{S,μ_S} which, in addition to equation (1.2), also satisfies the equations

$$\hat{S}^2 \Phi_{S,\mu_s} = S(S+1)\Phi_{S,\mu_s}, \quad \hat{S}_z \Phi_{S,\mu_s} = \mu_s \Phi_{S,\mu_s}. \tag{2.1}$$

For the simplest case N = 2 the function Φ which satisfies (1.1), (1.2) and (2.1) can be represented in the factorized form

$$\Phi_{S,\mu_S} = \Psi(\mathbf{r}_1,\mathbf{r}_2)u(\mu_1,\mu_2) \tag{2.2}$$

where the first factor Ψ is the coordinate part of Φ, and the second factor u is the spin part.

There are two possible values of S for the two-electron system: S = 0 and S = 1. In the first case u is antisymmetric with respect to transposition of its arguments and Ψ will be symmetrical with respect to transposition of its arguments. In the second case u will be symmetric and Ψ will be antisymmetric. However, if $N > 2$ then it is impossible to represent Φ by a simple product of two factors. Instead a linear combination of such products is needed. We will follow the treatment given by Fock (1940) (with some minor changes). To satisfy condition (1.1) irrespective of all others, it is sufficient to put

$$\Phi = \sum_{p_1 \ldots p_N} \delta_{p_1 \ldots p_N} \Psi(\mathbf{r}_{p_1} \ldots \mathbf{r}_{p_N}) u(\mu_{p_1} \ldots \mu_{p_N}). \tag{2.3}$$

Here $p_1 \ldots p_N$ are N unequal numbers obtained by a permutation

$$\begin{pmatrix} 1 & 2 & \ldots & N \\ p_1 & p_2 & \ldots & p_N \end{pmatrix}$$

where $\delta = 1$ for an even permutation and $\delta = -1$ for an odd one. The sum is extended over all possible permutations. Now let us divide the arguments of Ψ and u into two parts by a semicolon

$$\Psi(\mathbf{r}_1 \ldots \mathbf{r}_k;\ \mathbf{r}_{k+1} \ldots \mathbf{r}_N), \quad u(\mu_1 \ldots \mu_k;\ \mu_{k+1} \ldots \mu_N).$$

The number k is determined by the relation

$$k = \frac{N}{2} - S$$

where S is the total spin quantum number. It was shown by Fock that Φ satisfies the first of equations (2.1), if:

a) the function u is symmetrical with respect to interchange of any pair of arguments belonging to the same part (both on the left side or both on the right side of the sign ;)

$$(1 - P_{ij})u = 0 \tag{2.4}$$

$$i,j \leqslant k \text{ or } i,j > k.$$

Here P_{ij} denotes the permutation operator interchanging the two arguments i and j.

b) The function Ψ is antisymmetrical with respect to interchange of any pair of the argument belonging to the same part

$$(1 + P_{ij})\Psi = 0, \tag{2.5}$$

where $i,j \leq k$ or $i,j > k$.

c) The function Ψ possesses the property which was called by Fock "cyclic symmetry"

$$(1 - \sum_{j=k+1}^{N} P_{ij})\Psi = 0, \tag{2.6}$$

where $i \leq k$.

In order to ensure that Φ satisfies the second of the equations (2.1) one should construct a symmetrized linear combination of the products of N one-electron spin wave functions $u_{\frac{1}{2}}$ and $u_{-\frac{1}{2}}$. The simplest case is M = S. Then u is given by a single product

$$u = \overset{1}{u}_{-\frac{1}{2}} \cdots \overset{k}{u}_{-\frac{1}{2}}\overset{k+1}{u}_{\frac{1}{2}} \cdots \overset{N}{u}_{\frac{1}{2}}. \tag{2.7}$$

Note that in fact Ψ does not depend upon M.

So the problem of construction of the function Φ which satisfies (1.1), (1.2) and (2.1) is reduced to a more simple problem of constructing the function Ψ which satisfies (2.2), (2.5) and (2.6).

In the next section we will apply these general rules to the problem of electron-atom collisions.

4.3 REDUCTION OF THE MANY-CHANNEL PROBLEM

Let us consider first the simplest case: electron-hydrogen atom collisions. The singlet state of the system is described by a function $\Psi(\mathbf{r}_1;\mathbf{r}_2)$ (in the notation of the preceding section), the triplet state by a function $\Psi(;\mathbf{r}_1\mathbf{r}_2)$. These functions can be expanded in hydrogen atom eigenfunctions like

$$\Psi(;\mathbf{r}_1\mathbf{r}_2) = \sum_{\alpha}\psi_\alpha(\mathbf{r}_1)F_\alpha(\mathbf{r}_2). \tag{3.1}$$

Here the summation over atomic states includes an integration over the continuous spectrum.

The expression (3.1) at first sight does not possess the required symmetry with respect to interchange of $\mathbf{r}_1$ and $\mathbf{r}_2$ in an explicit form. But in fact (3.1) can be made symmetrical or antisymmetrical by choosing in a proper way the contour of integration in the integration over the continuous spectrum (Castillejo, Percival and Seaton, 1960).

An alternative representation for $\Psi(;r_1r_2)$ is

$$\Psi(;r_1r_2) = \sum_\alpha[\psi_\alpha(r_1)F^1_\alpha(r_2) - \psi_\alpha(r_2)F^1_\alpha(r_1)] \tag{3.2}$$

and for $\Psi(r_1;r_2)$ is

$$\Psi(r_1;r_2) = \sum_\alpha[\psi_\alpha(r_1)F^0_\alpha(r_2) + \psi_\alpha(r_2)F^0_\alpha(r_1)]. \tag{3.3}$$

Here each term in the sum possesses explicitly the required symmetry. The forms (3.2) and (3.3) are mostly used. To determine the functions F^0_α and F^1_α, we will use the relation

$$\int\Psi_\alpha(r_1)(H - E)\Psi dr_1 = 0 \tag{3.4}$$

where Ψ is given by (3.2) or (3.3).

If Ψ contains all hydrogen atom eigenfunctions, then Ψ is an exact solution of (2.2) and therefore (3.4) is satisfied. However, in actual calculations the approximate expression for Ψ is used containing a finite number of terms in the sum, rather than an exact solution of (2.2). In this case (2.2) is not satisfied but (3.4) still holds. It is important to note that (3.4) follows from the condition of an extremum of a functional

$$I = \int\Psi^*(H - E)\Psi dr_1 dr_2$$

with respect to arbitrary small variation of F_α (Demkov, 1958). The equations for $F^{S_t}_\alpha$ which follow from (3.4) after inserting (3.2) or (3.3) have the form

$$(\Delta + k^2_\alpha)F^{S_t}_\alpha = \sum_{\alpha'} V_{\alpha\alpha'}F^{S_t}_{\alpha'} + \sum_{\alpha'}\int W^{S_t}_{\alpha\alpha'}(r,r')F^{S_t}_{\alpha'}(r')dr' \tag{3.5}$$

where

$$V_{\alpha\alpha'} = 2\int \frac{\psi^*_\alpha(r')\psi_{\alpha'}(r')}{|r - r'|} dr' - \frac{2}{r}\delta_{\alpha\alpha'},$$

$$W^{S_t}_{\alpha\alpha'} = (-1)^{1-S_t} 2\psi^*_\alpha(r)[H - E]\psi_{\alpha'}(r'), \tag{3.6}$$

$W_{\alpha\alpha'}$ is called the exchange operator.

There is an essential difference in the behavior of $V_{\alpha\alpha'}$ and $W_{\alpha\alpha'}$ as a function of r at large r. $W_{\alpha\alpha'}$ decreases exponentially while $V_{\alpha\alpha'}$ decreases as r^{-n}, the power of r depending on angular momentum quantum numbers ℓ and ℓ'. The slowest decrease (in the case of neutral atoms) is r^{-2}. However, there is an exception in the case $\ell = \ell' = 0$ when $V_{\alpha\alpha'}$ also decreases exponentially.

In addition to satisfying equation (3.5), the asymptotic behavior of $F_\alpha^{S_t}$ must be specified:

$$F_\alpha^{S_t} \sim e^{i\mathbf{k}_0\cdot\mathbf{r}}\delta_{\alpha\alpha^0} + \frac{e^{ik_\alpha r}}{r}\langle\alpha|A^{S_t}|\alpha^0\rangle. \tag{3.7}$$

Note that $F_\alpha^{S_t}$ depends also upon the index α^0 so it would be more consistent to write $F_{\alpha\alpha^0}^{S_t}$.

The relations (3.5) - (3.7) form a many-channel problem, the simplest case of which was considered in Chapter 3. It should be noted that the number of equations in the system (3.5) is always infinite (provided the number of terms in the sum (3.2) or (3.3) is infinite) even if there is only one channel open.

Now we consider a more complicated case: electron-helium atom collisions. Let the atom be initially in its ground state (which is a singlet state). The total spin of "electron + atom" system is equal to 1/2. The corresponding coordinate wave function is then $\Psi(\mathbf{r}_1;\mathbf{r}_2\mathbf{r}_3)$. It should be constructed like (3.2) and (3.3) from the atomic wave functions and that of the scattered electron. There are two kinds of atomic wave functions: $\psi(\mathbf{r}_1;\mathbf{r}_2)$ for singlet states and $\psi(;\mathbf{r}_1\mathbf{r}_2)$ for triplet states which are involved. It is easy to verify that the expressions

$$\psi(\mathbf{r}_1;\mathbf{r}_2)F(\mathbf{r}_3) - \psi(\mathbf{r}_1;\mathbf{r}_3)F(\mathbf{r}_2)$$

and

$$2\psi(;\mathbf{r}_3\mathbf{r}_2)F(\mathbf{r}_1)+\psi(;\mathbf{r}_1\mathbf{r}_2)F(\mathbf{r}_3) - \psi(;\mathbf{r}_1\mathbf{r}_3)F(\mathbf{r}_2)$$

satisfy relations (2.5) - (2.6) for the function

$$\Psi(\mathbf{r}_1;\mathbf{r}_2\mathbf{r}_3).$$

Taking this into account, we put

$$\Psi(\mathbf{r}_1;\mathbf{r}_2\mathbf{r}_3)$$

$$= C_0\sum_\beta[\psi_\beta(\mathbf{r}_1;\mathbf{r}_2)F^{1/2}_{\beta\alpha^0}(\mathbf{r}_3) - \psi_\beta(\mathbf{r}_1;\mathbf{r}_3)F^{1/2}_{\beta\alpha^0}(\mathbf{r}_2)] \tag{3.8}$$

$$+ C_1\sum_\gamma[2\psi_\gamma(;\mathbf{r}_3\mathbf{r}_2)F^{1/2}_{\gamma\alpha^0}(\mathbf{r}_1) + \psi_\gamma(;\mathbf{r}_1\mathbf{r}_2)F^{1/2}_{\gamma\alpha^0}(\mathbf{r}_3) - \psi_\gamma(;\mathbf{r}_1\mathbf{r}_3)F^{1/2}_{\gamma\alpha^0}(\mathbf{r}_2)].$$

Here the index β refers to a singlet atomic state ($S = 0$) and γ to a triplet one ($S = 1$). In the functions $F^{1/2}_{\beta\alpha^0}$ and $F^{1/2}_{\gamma\alpha^0}$ we include an initial state index α^0 upon which these functions depend implicitly.

The coefficients C_0 and C_1 will be determined from the normalization condition of the spin part of the total wave function Φ describing the system "electron + helium atom." With this aim, let us insert (3.8) and (2.7) into (2.3). The resulting expression for Φ can be represented in the form

$$\Phi = (1-P_{12}-P_{13})\{\sum_{\beta} F^{1/2}_{\beta\alpha_0}(r_1)\psi_\beta(r_2;r_3)C_0 u^1_{\frac{1}{2}}[u^2_{\frac{1}{2}}u^3_{-\frac{1}{2}} - u^3_{\frac{1}{2}}u^2_{-\frac{1}{2}}]$$

$$+ \sum_{\gamma} F^{1/2}_{\gamma\alpha_0}(r_1)\psi_\gamma(;r_2r_3)C_1[2u^1_{-\frac{1}{2}}u^2_{\frac{1}{2}}u^3_{\frac{1}{2}} - u^1_{\frac{1}{2}}(u^2_{\frac{1}{2}}u^3_{-\frac{1}{2}} + u^3_{\frac{1}{2}}u^2_{-\frac{1}{2}})]\}.$$

In order to get the spin functions mormalized to 1, one should put

$$C_o = 1/\sqrt{2}, \quad C_1 = 1/\sqrt{6}.$$

Inserting (3.8) in the relation

$$\int\psi_\alpha(H-E)\Psi dr_1 dr_2 = 0$$

where α denotes β or γ, we get the equations for F

$$(\Delta + k^2_\alpha)F^{1/2}_{\alpha\alpha_0} = \sum_{\alpha'} V_{\alpha\alpha'}F^{1/2}_{\alpha'\alpha_0} + \sum_{\alpha'}\int W_{\alpha\alpha'}F^{1/2}_{\alpha'\alpha_0}dr' \tag{3.9}$$

where

$$V_{\alpha\alpha'} = 2\int \frac{\rho_{\alpha\alpha'}(r')}{|r-r'|}dr' - \frac{4}{r}\delta_{\alpha\alpha'}. \tag{3.10}$$

For $\alpha = \beta$:

$$\rho_{\beta\beta'}(r) = \int\psi^*_\beta(r;r'')\psi_{\beta'}(r;r'')dr''. \tag{3.11}$$

For $\alpha = \gamma$:

$$\rho_{\gamma\gamma'}(r) = \int\psi^*_\gamma(;rr'')\psi_{\gamma'}(;rr'')dr''.$$

The kernels of the exchange operator are:

For $\alpha = \beta$, $\alpha' = \beta'$:

$$W_{\beta\beta'} = -\int\psi^*_\beta(r;r'')(H-E)\psi_{\beta'}(r'';r')dr''.$$

For $\alpha = \gamma$, $\alpha' = \beta'$:

$$W_{\gamma\beta'} = -\sqrt{3}\int\psi^*_\gamma(;r'r'')(H-E)\psi_{\beta'}(r'';r')dr''.$$

For $\alpha = \beta$, $\alpha' = \gamma'$

$$W_{\beta\gamma'} = \sqrt{3}\int\psi^*_\beta(r;r'')(H-E)\psi_{\gamma'}(;r'';r')dr''.$$

For $\alpha = \gamma$, $\alpha' = \gamma'$

$$W_{\gamma\gamma'} = \int \psi_\gamma^*(;\mathbf{r}\mathbf{r}'')(H-E)\psi_{\gamma'}(;\mathbf{r}'',\mathbf{r}')d\mathbf{r}''. \tag{3.12}$$

Finally, the asymptotic expression for $F_{\alpha\alpha^0}$:

$$F^{1/2}_{\alpha\alpha^0} \sim e^{i\mathbf{k}_0\cdot\mathbf{r}}\delta_{\alpha\alpha^0} + \frac{e^{ik_\alpha r}}{r}\langle\alpha|A^{1/2}|\alpha^0\rangle.$$

Expressions like (3.8) and the corresponding standard equation like (3.9) can be constructed for the general case of an arbitrary atom as a target. The procedure is rather tedious and will not be reproduced here. The detailed derivation is given in another place (Drukarev, 1965).

4.4 PARTIAL WAVES: THE TOTAL ANGULAR MOMENTUM REPRESENTATION

Like the partial wave expansion in the case of potential scattering, we can expand the functions $F^{S_t}_{\alpha\alpha^0}$. For brevity we drop the total spin index S_t and initial state index α^0 on F. Then

$$F_\alpha(\mathbf{r}) = \sum_\ell \sqrt{4\pi(2\ell+1)}\; i^\ell \frac{f_\alpha^{\ell m}(r)}{r} Y_{\ell m}(\theta,\phi). \tag{4.1}$$

The electronic angular momentum projection quantum number m is fixed because of the conservation of the total angular momentum projection

$$\mu + m = \text{const}.$$

For this reason the summation in (4.1) goes over ℓ only. The radial wave functions satisfy the equation

$$\left(\frac{d^2}{dr^2} + k_\alpha^2 - \frac{\ell(\ell+1)}{r^2}\right) f_\alpha^{\ell m} = \sum_{\alpha'\ell'm'}\left(V_{\alpha\alpha'}^{\ell m\ell'm'} f_{\alpha'}^{\ell'm'} + \int_0^\infty W_{\alpha\alpha'}^{\ell m\ell'm'} f_{\alpha'}^{\ell'm'} dr'\right) \tag{4.2}$$

where

$$V_{\alpha\alpha'}^{\ell m\ell'm'} = \int Y_{\ell m}^* V_{\alpha\alpha'}(\mathbf{r}) Y_{\ell'm'} d\Omega,$$
$$W_{\alpha\alpha'}^{\ell m\ell'm'} = rr' \int Y_{\ell m}^*(\theta,\phi) W_{\alpha\alpha'}(\mathbf{r}\mathbf{r}') Y_{\ell'm'}(\theta'\phi') d\Omega d\Omega'. \tag{4.3}$$

The asymptotic expression for $f_\alpha^{\ell m}$ has the form

$$f_\alpha^{\ell m} \sim \frac{1}{k_0}\sin\left(k_0 r - \frac{\ell_0\pi}{2}\right)\delta_{\alpha\alpha^0} + e^{i(k_\alpha r - \frac{\ell\pi}{2})}\langle\alpha\ell m|A^{S_t}|\alpha^0\ell^0 m^0\rangle. \tag{4.4}$$

The functions ψ_α and $f_\alpha^{\ell m}$ describe the state of the system in the $LM\ell m$ representation. It is advantageous to transform to the $L\ell L_t M_t$ representation where L_t is the total angular momentum of the system "electron + atom," and M_t is its projection. The relation between the quantities in both representations has the form

$$f_\Gamma^{L_t M_t} = \sum_{mM} C_{LM\ell m}^{L_t M_t} C_{L^0 M^0 \ell^0 m^0}^{L_t M_t} f_\alpha^{\ell m},$$

$$V_{\Gamma\Gamma'}^{L_t M_t} = \sum_{mMm'M'} C_{LM\ell m}^{L_t M_t} C_{L'M'\ell'm'}^{L_t M_t} V_{\alpha\alpha'}^{\ell m \ell' m'}, \tag{4.5}$$

$$W_{\Gamma\Gamma'}^{L_t M_t} = \sum_{mMm'M'} C_{LM\ell m}^{L_t M_t} C_{L'M'\ell'm'}^{L_t M_t} W_{\alpha\alpha'}^{\ell m \ell' m'},$$

where $\Gamma = (nL\ell S)$.

The functions $f_\Gamma^{L_t M_t}$ satisfy the system of equations

$$\left(\frac{d^2}{dr^2} + k_\Gamma^2 - \frac{\ell(\ell+1)}{r^2}\right) f_\Gamma^{L_t M_t} = \sum_{\Gamma'}\left(V_{\Gamma\Gamma'}^{L_t M_t} f_{\Gamma'}^{L_t M_t} + \int_o^\infty W_{\Gamma\Gamma'}^{L_t M_t} f_{\Gamma'}^{L_t M_t} dr'\right) \tag{4.6}$$

with the asymptotic form

$$f_\Gamma^{L_t M_t} \sim \frac{1}{k_0} \sin\left(k_0 r - \frac{\ell_0 \pi}{2}\right)\delta_{\Gamma\Gamma^0} + e^{i\left(k_\Gamma r - \frac{\ell\pi}{2}\right)} \langle \Gamma | A^{L_t S_t} | \Gamma^0 \rangle. \tag{4.7}$$

Here $\langle\Gamma|A^{L_t S_t}|\Gamma^0\rangle$ is the amplitude in the total angular momentum representation. To formulate its relation to the amplitude $\langle\alpha\ell m| |A^{S_t}|\alpha^0\ell^0 m^0\rangle$, we will write down nLMS instead of α:

$$\langle nL\ell S|A^{L_t S_t}|n^0 L^0 \ell^0 S^0\rangle = \sum_{MmM^0 m^0} C_{LM\ell m}^{L_t M_t} C_{L^0 M^0 \ell^0 m^0}^{L_t M_t} \langle nLM\ell mS|A^{S_t}|n^0 L^0 M^0 \ell^0 m^0 S^0\rangle. \tag{4.8}$$

The inverse relation has the form

$$\langle nLM\ell mS|A^{S_t}|n^0 L^0 M^0 \ell^0 m^0 S^0\rangle = \sum_{L_t} C_{LM\ell m}^{L_t M_t} C_{L^0 M^0 \ell^0 m^0}^{L_t M_t} \langle nL\ell S|A^{L_t S_t}|n^0 L^0 \ell^0 S^0\rangle. \tag{4.9}$$

Note that the amplitude in the total angular momentum representation does not depend on the total angular momentum projection M_t.

The system of equations (4.6) in contrast to (4.2) can be decomposed into independent blocks for each L_t separately. This is the advantage of the total angular momentum representation. The reason for this is the angular momentum addition rule. for given L_t and L the quantum number ℓ is limited by the condition $|L_t - L| \leqslant \ell \leqslant L_t + L$. Moreover, the equations for even ℓ are separated from the equations for odd ℓ.

4.5 THE JOST MATRIX AND RELATED MATRICES

Like the case of potential scattering, we consider the functions $R_{\alpha\alpha^0}$ which satisfy the system of equations

$$\left(\frac{d^2}{dr^2} + k_\alpha^2 - \frac{\ell(\ell+1)}{r^2}\right) R_{\alpha\alpha^0} = \sum_{\alpha'} [V_{\alpha\alpha'} R_{\alpha'\alpha^0} + \int_0^\infty W_{\alpha\alpha'} R_{\alpha'\alpha^0} dr'] \quad (5.1)$$

and the following conditions at the origin:

$$R_{\alpha\alpha^0}(0) = 0,$$

$$\frac{d}{dr}(R_{\alpha\alpha^0} r^{-\ell})_{r=0} = \frac{k_\alpha{}^\ell}{(2\ell+1)!!} \delta_{\alpha\alpha^0}. \quad (5.2)$$

The asymptotical expression for $R_{\alpha\alpha^0}$ can be represented in the form

$$R_{\alpha\alpha^0} \sim \frac{1}{2ik_\alpha}\left(e^{i(k_\alpha r - \frac{\ell\pi}{2})} I^*_{\alpha\alpha^0}(k_\alpha) - e^{-i(k_\alpha r - \frac{\ell\pi}{2})} I_{\alpha\alpha^0}(k_\alpha)\right) \quad (5.3)$$

for the open channels where $I_{\alpha\alpha^0}(k_\alpha)$ is the Jost matrix. For the closed channels $R_{\alpha\alpha^0}(r) \to 0$ when $r \to \infty$.

Note that $R_{\alpha\alpha^0}$ goes to zero as $r \to \infty$ for the closed channels like some inverse power of r due to the long-range behavior of $V_{\alpha\alpha^0}(r)$.

Using a concise matrix notation, we rewrite (5.3) as

$$R = (2ik)^{-1}\left(e^{i(kr - \frac{\ell\pi}{2})} I^* - e^{-i(kr - \frac{\ell\pi}{2})} I\right). \quad (5.4)$$

The functions f and R are related by the expression

$$R = fI. \quad (5.5)$$

The asymptotic expression for f is

$$f \sim k^{-1}\sin(kr - \frac{\ell\pi}{2}) + e^{i(kr - \frac{\ell\pi}{2})} A. \quad (5.6)$$

Using this expression and taking into account (5.4), we have

$$A = (2ik)^{-1}(I^{*}I^{-1} - 1). \tag{5.7}$$

The Jost matrix can be represented in the form

$$I = k^{\ell}Ck^{-\ell} + ik^{\ell}Bk^{\ell+1} \tag{5.8}$$

where C and B are real matrices depending upon k_α^2.

Note that by means of (1.6) all k_α^2 can be expressed in terms of k_o^2. Putting $C = -BM$, we get

$$I = -k^{\ell}B(M - ik^{2\ell+1})k^{-\ell} \tag{5.9}$$

which leads to

$$A = k^{\ell}(M - ik^{2\ell+1})k^{\ell}. \tag{5.10}$$

The unitary S matrix is related to A in the following way:

$$S = 1 + 2ik^{\frac{1}{2}}Ak^{\frac{1}{2}}. \tag{5.11}$$

4.6 EFFECTS OF CLOSED CHANNELS

It was mentioned in the Introduction that the response of an atom to a slow moving electron can be pictured as a polarization which leads to an addition to the interaction energy of a term $-\alpha/r^4$ at large distances. It is interesting to see how this r^{-4} term can be extracted from equation (3.5). Here we follow the treatment given by Castillejo, Percival and Seaton (1960). Let us consider (3.5) at large distances and drop all interactions except the most slowly decreasing. Then the most important equations in the system (3.5) will be

$$\Delta F_o + k_o^2 F_o = \sum_\alpha V_{o\alpha}F_\alpha, \qquad \Delta F_\alpha + k_\alpha^2 F_\alpha = V_{\alpha o}F_o. \tag{6.1}$$

Here we have dropped some indices which do not matter. At low energy all $k_\alpha^2 < 0$ ($\alpha \neq 0$). In this case the leading term in the left side of the second equation (6.1) will be $k_\alpha^2 F_\alpha$ so

$$F_\alpha \simeq \frac{V_{\alpha o}}{k_\alpha^2}F_o = \frac{V_{\alpha o}F_o}{2(E_o - E_\alpha) + k_o^2}. \tag{6.2}$$

Inserting (6.2) in the first equation of (6.1) and taking into account that $V_{\alpha o}$ decreases as r^{-2}, we have

$$\Delta F_0 + k_0^2 F_0 = \frac{\text{const}}{r^4} F_0. \qquad (6.3)$$

A more detailed consideration in the cited paper reveals that the "const" in the right side of (6.3) is in fact proportional to the atomic polarizability in the limit $k_0^2 \to 0$.

Near the excitation threshold the effect of closed channels leads to resonances similar to that discussed in Chapter 3. Of special interest is the Coulomb-like series of resonances in the case of electron-positive ion collision. These resonances were considered by Seaton (1969) using the multichannel quantum-defect theory.

The basic expression for the reactance matrix in the simplest case of a two-channel system was considered in Section 3.5 (see expression 5.9). For the general case of many-channel systems, the expression for K-matrix can be written in the form

$$K = K_{oo} - K_{oc}(K_{cc} + \tan\frac{\pi Z}{\kappa})^{-1} K_{co}$$

where the elements of the submatrices K_{oo},K_{oc}, K_{co} and K_{cc} are obtained by analytical continuation of the K-matrix from above the threshold to energies below the threshold and by partitioning the analytical continuation of K as

$$\begin{pmatrix} K_{oo} & K_{oc} \\ K_{co} & K_{cc} \end{pmatrix}$$

(o refers to "open", c refers to "closed").

Another approach was used by Presnyakov and Urnov (1975). Here we will demonstrate this approach on a simplified two-channel system. We will consider only the $\ell = 0$ partial wave. In the diagonal elements of the interaction matrix we will drop all the terms except the Coulomb field $-Z_0/r$ where Z_0 is the charge of the ion. The only nondiagonal interaction matrix element will be denoted by V. It is advantageous to use so-called Coulomb units in which $Z_0 r$ is replaced by r and k/Z_0 by k.

The equations for the two radial wave functions f_1 and f_0 will have the form

$$\frac{d^2 f_1}{dr^2} + (-\kappa^2 + \frac{1}{r}) f_1 = \frac{1}{Z_0} V f_0,$$
$$\frac{d^2 f_0}{dr^2} + (k_0^2 + \frac{1}{r}) f_0 = \frac{1}{Z_0} V f_1. \qquad (6.4)$$

It is important to note that V in the case of large Z_0 is almost independent of Z_0. Let us consider the first equation of (6.4). The formal solution of it can be written:

$$f_1(r) = \frac{1}{Z_0}\int_0^\infty Vg(i\kappa,r,r')f_0 dr' \tag{6.5}$$

where $g(i\kappa,r,r')$ is the Coulomb Green function. Presnyakov and Urnov have found an interesting decomposition of g:

$$g(i\kappa,r,r') = \frac{\pi\nu^3}{2a^2}\cot(\pi\nu)P_\nu(r)P_\nu(r') + g' \tag{6.6}$$

where $a = 1 - 1/Z_0$, $\nu = a/\kappa$ and P_ν are the Coulomb field wave functions. In the case where ν is an integral positive number, P_ν coincides with the wave function of a bound state in the Coulomb field. The most important first term in (6.6) has a factorized form. The $\pi\nu^3/2a^2 \cot\pi\nu$ factor which is responsible for the resonances does not depend on r and therefore it is easy to separate this factor in the calculations which follow. The second term g' is a smooth function of energy. Inserting (6.6) into (6.5), we get

$$f_1(r) = \frac{1}{Z_0}\frac{\pi\nu^3}{2a^2}\cot\pi\nu P_\nu(r)\int_0^\infty P_\nu(r')Vf_0 dr' + \frac{1}{Z_0}\int_0^\infty f_0 Vg'dr', \tag{6.7}$$

and after inserting (6.7) into (6.4), we get an equation for f

$$\frac{d^2f_0}{dr^2} + (k_0^2 + \frac{1}{r})f_0 = \frac{1}{Z_0^2}\frac{\pi\nu^3}{2a^2}\cot\pi\nu VP_\nu(r)\int_0^\infty P_\nu Vf_0 dr' + \frac{1}{Z_0^2}V\int_0^\infty Vg'f_0 dr'. \tag{6.8}$$

Let us perform some transformations in (6.8). Denote

$$\frac{\pi\nu^3}{2a^2}\cot\pi\nu\int_0^\infty P_\nu Vf_0 dr' = \Lambda, \tag{6.9}$$

and put

$$f_0 = \Lambda x + y \tag{6.10}$$

where x and y satisfy the equations

$$\begin{aligned} &\frac{d^2x}{dr^2} + (k_0^2 + \frac{1}{r})x - \frac{1}{Z_0^2}\int_0^\infty Vg'xdr' = \frac{1}{Z_0^2}VP_\nu, \\ &\frac{d^2y}{dr^2} + (k_0^2 + \frac{1}{r})y - \frac{1}{Z_o^2}\int_o^\infty Vg'ydr' = 0. \end{aligned} \tag{6.11}$$

Inserting (6.10) into (6.9) we get an equation for Λ, the solution of which is

$$\Lambda = \frac{\int_0^\infty P_\nu Vydr'}{Z_0^2\frac{2a^2}{\pi\nu^3}\tan\pi\nu - \int_0^\infty P_\nu Vxdr'}. \tag{6.12}$$

So the problem is reduced to the solution of Equation (6.11) for x and y. We will require that x and y are regular at the origin and have the following asymptotic expressions:

$$\begin{aligned} y &\sim \sin(k_0 r - \frac{1}{k_0} \ln 2k_0 r + \eta_0 + \alpha), \\ x &\sim \sin(k_0 r - \frac{1}{k_0} \ln 2k_0 r + \eta_0 + \beta) \end{aligned} \tag{6.13}$$

where η_0 is the Coulomb phase and α, β are the additional phase shifts.

Then for f_0 we get the following asymptotic expression

$$f_0 \sim C \sin(k_0 r - \frac{1}{k_0} \ln 2k_0 r + \eta_0 + \delta) \tag{6.14}$$

where

$$\tan = \frac{\sin\alpha + \Lambda\sin\beta}{\cos\alpha + \Lambda\cos\beta} . \tag{6.15}$$

The additional phase shift δ represents the influence of the closed channel. Due to the presence of $\tan \pi\nu$ in (6.12), δ becomes $\pi/2$ at some values of energy below the threshold which is indeed the resonance effect we are looking for. If $Z_0 >> 1$, then $1/Z_0$ is a small parameter. The phases α and β become small. In the first approximation the resonances occur at $\Lambda = -1$ or

$$\cot \frac{\pi}{\kappa} = \frac{Z_0^2 \kappa^3}{2\pi \int_0^\infty P_\nu V(x - y) dr} . \tag{6.16}$$

This condition is similar to (5.6) of Chapter 3. The only difference is that instead of the interaction V in Chapter 3 boundary conditions have been used. Presnyakov and Urnov developed a method of solution of the many-channel problem equations by means of successive approximations in terms of a small parameter $1/Z_0$ for an arbitrary number of closed and open channels.

The general conclusion which follows from this work is that at large Z_0 the coupling between open channels is rather a small correction in comparison with the main effect of the closed channels.

The averaged-over-small-energy-interval cross section for elastic scattering has steps of the thresholds similar to those discussed in Section 3.5. An averaged excitation cross section of the energy level number n has steps at the excitation thresholds of the energy level numbers n + 1, n + 2 ...

4.7 CORRELATION AND POLARIZATION PHENOMENA

The density matrix formalism outlined in Chapter 2 gives the most general and exhaustive characteristics of an atomic ensemble selected in a given experiment. Let us consider the electron-photon coincidence experiment in which the atom is excited by an electron and subsequently emits a photon which is observed in coincidence with the scattered electron. The following assumptions will be made: a) excitation and emission are independent processes. The atomic mean life time is supposed to be long enough; b) both electrons and atoms are unpolarized before the collision; c) the spin-orbit interaction is neglected; and d) the initial atoms are in their ground state having zero angular momentum.

To construct the relevant density matrix it is instructive to look at the spin density matrix after the collision [Chapter 2, equation (3.4)]. Following this example, we should write in the present case also

$$\rho = \frac{A\rho_o A^+}{Sp(A\rho_o A^+)} \qquad (7.1)$$

with the only difference that both ρ_o and A contain, besides the spin projections of the atom μ_s and electron μ, also the angular momentum projections M.

However, Blum and Kleinpoppen (1978) (hereafter BK) and other authors dropped the normalization factor $Sp(A\rho_o A^+)^{-1}$ introduced to make $Sp\rho = 1$, and used the unnormalized density matrix

$$\rho = A\rho_o A^+. \qquad (7.2)$$

To reduce the general density matrix $\langle\mu\mu_s M|\rho|\mu'\mu_s' M'\rangle$ to the matrix for the angular momentum projections M only, $\langle M|\rho|M'\rangle$, in the situation when there are no spin measurements, one should put $\mu' = \mu$, $\mu_s' = \mu_s$ and take the sum over all μ, μ_s

$$\langle M|\rho|M'\rangle = \Sigma_{\mu\mu_s} \langle\mu\mu_s M|\rho|\mu\mu_s M'\rangle . \qquad (7.3)$$

Now the initial atomic state is supposed to have zero angular momentum so that the initial density matrix depends in fact only upon the spin variables $\mu^0\mu_s^0$. In the case of unpolarized electrons and atoms the initial density matrix is a unity matrix up to a normalization factor:

$$\rho_o = \frac{1}{2(2S^0+1)} .$$

Inserting (7.3) into (7.2), we get

$$\langle M|\rho|M'\rangle = \frac{1}{2(2S^0+1)} \sum_{\mu\mu_s} \langle LMS\mu_s\mu|AA^+|L'M'S'\mu_s\mu\rangle . \qquad (7.4)$$

We rewrite (7.4) in a more concise form:

$$\langle M|\rho|M'\rangle = \langle A_M A^*_{M'}\rangle. \tag{7.5}$$

It follows from (7.5) that

$$Sp\rho = \langle\sum_M |A_M|^2\rangle = \sigma. \tag{7.6}$$

In some cases it is advantageous to use the total spin representation in which

$$\langle M|\rho|M'\rangle = \sum_{S_t} (2S_t + 1)A_M^{S_t}(A_{M'}^{S_t})^*. \tag{7.7}$$

The density matrix obeys a symmetry relation which is a consequence of (1.9)

$$\langle M'|\rho|M\rangle = (-1)^{M' + M}\langle -M'|\rho|-M\rangle. \tag{7.8}$$

There are two special cases of above general relations. The first is a transition between singlet states of atoms ($S = S_0 = 0$). Having neglected the spin-orbit interaction, we will get in this case an amplitude independent of the scattered electron's spin projection. Then the density matrix does not depend upon spin variables and is given by the expression

$$\langle M|\rho|M'\rangle = A_M A^*_{M'}. \tag{7.9}$$

We have here a pure state (in the sense explained in Chapter 2). The second special case is a singlet-triplet transition ($S_0 = 0$, $S = 1$). Because of spin conservation in (7.7), only the total spin $S_\gamma = 1/2$ will contribute. As a result we have again the density matrix in a factorized form

$$\langle M|\rho|M'\rangle = A_M^{1/2}(A_{M'}^{1/2})^*. \tag{7.10}$$

Now let us count the number of independent real parameters needed to determine the density matrix.

It was shown in Chapter 2 that a normalized N x N density matrix has $N^2 - 1$ real parameters for a mixed state and $2N - 2$ for a pure state. An unnormalized density matrix has correspondingly N^2 and $2N - 1$ real independent parameters. In the case under consideration $N = 2L + 1$. However, we should take into account the relation (7.8). Because of this relation only half of the $N^2 - 1$ matrix elements $\langle M|\rho|M'\rangle$ with either M or $M' \neq 0$ will be independent. Adding the element $\langle 0|\rho|0\rangle$ which is unaffected by (7.8), we have $\frac{N^2 - 1}{2} + 1$ independent real parameters for a general mixed state or $2L^2 + 2L + 1$ parameters in terms of L.

For a pure state we have a similar reduction: $\frac{(2N-1)-1}{2}+1$ parameters instead of 2N - 1 which is equal to N or 2L + 1 in terms of L.

As an example, let us consider a P state, L = 1. The number of independent real parameters will be 5 for a mixed state, and 3 for a pure one. In the special case of a singlet-triplet transition, because of the factorization (7.10), we have 3 parameters also.

A convenient parametrization for the factorized density matrix (7.9) or (7.10) is mentioned in BK. A_0 is taken to be real and $A_{\pm 1}$ are represented in the form $\pm |A_1| e^{i\chi}$. Then the three parameters are:

$$\sigma,\ \chi \text{ and } \lambda = \frac{|A_0|^2}{\sigma} \tag{7.11}$$

[where σ is defined in (7.6)].

An alternative parametrization of the density matrix is possible in terms of the tensor operators (Chapter 2). We reproduce here relations between the mean values of tensor operators and the parameters σ, χ, λ for a ^1S - ^1P transition (BK):

$$\begin{aligned} \langle T_{00}\rangle &= \frac{\sigma}{\sqrt{3}}, \quad \langle T_{11}\rangle = -i\sigma\sqrt{\lambda(1-N)}\sin\chi, \\ \langle T_{22}\rangle &= 1/2\sigma(\lambda-1), \\ \langle T_{21}\rangle &= -\sigma\sqrt{\lambda(1-\lambda)}\cos\chi, \\ \langle T_{20}\rangle &= \frac{1}{\sqrt{6}}\sigma(1-3\lambda). \end{aligned} \tag{7.12}$$

It follows from the discussion in Chapter 2 that the value $\langle T_{11}\rangle$ characterizes the polarization, or orientation of the system, and $\langle T_{20}\rangle$, $\langle T_{21}\rangle$ and $\langle T_{22}\rangle$ characterize the alignment. To establish the connection with observable quantities, we should consider the description of light polarization by means of a density matrix. Let $\mathbf{e}_1$ and $\mathbf{e}_2$ be two orthogonal unit vectors in the plane perpendicular to the propagation direction of the photon $\mathbf{n}$. The polarization state can be characterized by a density matrix

$$\rho = \frac{1}{2}(1 + \eta_1\sigma_x + \eta_2\sigma_y + \eta_3\sigma_z). \tag{7.13}$$

However, the indices x, y, z do not refer to any coordinate system in real space. They are introduced merely to denote the three Pauli matrices. Positive values of the parameter η_3 describe the degree of linear polarization along $\mathbf{e}_1$ and negative values of η_3 are that along $\mathbf{e}_2$. The values of η_1 describe the degree of linear polarization along the directions $\pm$ 45° to $\mathbf{e}_1$ (+ for positive η_1,

- for negative η_1); η_2 describes the circular polarization. The four quantities: J, the light intensity emitted in the direction **n** and averaged over polarization, $J\eta_1$, $J\eta_2$, $J\eta_3$, are called the Stokes parameters.

The final step is to find a relation between the Stokes parameters and the parameters defined by the density matrix $\langle M|\rho|M'\rangle$. Again we consider here only the simplest case of $^1S - P^1$ excitation referring to BK for the more complicated case (including taking account of fine structure of atomic levels). We reproduce here the relation between σ, χ, λ and the Stokes parameters. Before writing them down, let us specify the coordinate system relevant to experimental conditions.

Let the z-axis of the coordinate system coincide with the direction of the incident electron, the momentum of the scattered electron lies in the xz-plane (the scattering plane), and the propagation direction of the photon **n** is specified by the polar angles θ, ϕ. Now we write down the expressions for J and $J\eta_2$:

$$J = J_o \frac{\sigma}{3}\Big[\frac{1-\lambda}{2}(1-\sin^2\theta\cos2\phi+\cos^2\theta)+\lambda\sin^2\theta + \sqrt{\lambda(1-\lambda)}\cos\chi\sin2\theta\cos\phi\Big], \tag{7.14}$$

$$J\eta_2 = J_o \frac{2}{3}\sigma\sqrt{\lambda(1-\lambda)}\sin\chi\sin\theta\sin\phi. \tag{7.15}$$

Here we indicate explicitly the dependence upon λ, σ and χ. All other factors are absorbed in J_o.

If the photons are detected in the y direction ($\theta = \pi/2$, $\phi = \pi/2$), then

$$\eta_2 = -2\sqrt{\lambda(1-\lambda)}\sin\chi. \tag{7.16}$$

If the photons are detected in the scattering plane, it can be shown that they are completely linearly polarized, $\eta_3 = 1$.

In concluding this section we will consider an example of a spin-polarization phenomenon in an inelastic collision, under conditions where the atomic orbital angular momentum is not involved, namely, $1^1S - 2^3S$ transition in He, induced by the electron impact. To simplify the problem, we will neglect completely the spin-orbit interaction.

The main problem is to construct an operator acting on spin variables which describe the transition.

Let us denote by index 1 the spin variables of the scattered electron, and by index 2 and 3 the spin variables of atomic electrons. The state vector which corresponds to the system "He atom in singlet state + electron" can be expressed in the form

$$u_i = |\tfrac{1}{2}>_1 \cdot \frac{1}{\sqrt{2}}(|\tfrac{1}{2}>_2|-\tfrac{1}{2}>_3 - |-\tfrac{1}{2}>_2|\tfrac{1}{2}>_3). \qquad (7.17)$$

The state vector of the system "He atom in triplet state + electron" corresponding to the same total spin and its projection as the vector (7.17) is equal to

$$u_f = \frac{1}{\sqrt{6}}[2|-\tfrac{1}{2}>_1|\tfrac{1}{2}>_2|\tfrac{1}{2}>_3 - |\tfrac{1}{2}>_1(|\tfrac{1}{2}>_2|-\tfrac{1}{2}>_3 + |-\tfrac{1}{2}>_2|\tfrac{1}{2}>_3)]. \qquad (7.18)$$

It is easy to verify that the operator

$$M = \frac{1}{2\sqrt{3}}\boldsymbol{\sigma}_1(\boldsymbol{\sigma}_2 - \boldsymbol{\sigma}_3) \qquad (7.19)$$

transforms u_i into u_f and u_f into u_i:

$$Mu_i = u_f, \quad Mu_f = u_i. \qquad (7.20)$$

Taking this into account we represent the amplitude operator in the form

$$\hat{A} = AM \qquad (7.21)$$

where A is the numerical value of the scattering amplitude. Next we must cónstruct the density matrix corresponding to the initial state. It should be a product of density matrices of the electronic beam and of the He atoms (in a singlet state). The electronic beam matrix is equal to $1/2(1+\mathbf{P}\cdot\boldsymbol{\sigma})$ and the atomic matrix can be represented in the form $1/4(1-\boldsymbol{\sigma}_2\cdot\boldsymbol{\sigma}_3)$, so that the total density matrix is

$$\rho = \frac{1}{8}(1-\boldsymbol{\sigma}_2\cdot\boldsymbol{\sigma}_3)(1+\mathbf{P}\cdot\boldsymbol{\sigma}_1). \qquad (7.22)$$

The density matrix after the collision is

$$\rho' = \frac{A\rho A^+}{Sp(A\rho A^+)}. \qquad (7.23)$$

It follows that the electronic polarization is

$$\mathbf{P}_1' = Sp(\boldsymbol{\sigma}_1\rho') = -\frac{1}{3}\mathbf{P}. \qquad (7.24)$$

The polarization of the triplet helium atoms is equal to

$$\mathbf{P}_2' = Sp(\boldsymbol{\sigma}_2\rho') = \frac{2}{3}\mathbf{P}. \qquad (7.25)$$

5
Approximate Methods for Electron-Atom Collisions: Cross Section Calculations

In this chapter some of the approximate methods for cross section calculations will be considered. We do not intend to consider all existing methods, but only those which are mostly used and can be applied to a variety of particular cases. We will compare the results of calculations by various methods and with the experimental data.

5.1 METHODS FOR THE APPROXIMATE SOLUTION OF THE MANY-CHANNEL PROBLEM

1. The Close Coupling Method

The idea of the close coupling method in its simplest form can be demonstrated for the example of e - H scattering. It is to replace the infinite number of terms in (3.2) or (3.3) of Chapter 4 by a finite number, taking into account the most important open channels and some of the closed:

$$\Psi = \sum_{n=1}^{N} [\psi_n(r_1)F_n(r_2) \pm \psi_n(r_2)F_n(r_1)]. \qquad (1.1)$$

The system of equations for the functions F_n, which follows after inserting (1.1) in the relation (3.4) of Chapter 4, will be a finite set instead of an infinite set as in (3.5) of Chapter 4. Separating the angular variables we obtain equations for the partial waves like (4.2) or (4.6) of Chapter 4, but again for the finite number of target states instead of an infinite number.

The partial wave amplitude for a given transition which is the result of the calculations is an intermediate quantity. It is the weighted sum of partial cross sections which should be compared with experiment rather than some particular partial cross section. Nevertheless, the study of the energy dependence of each partial cross section separately can give important information because the resonances occur in some particular partial cross section.

It seems obvious that, the more channels which are taken into account in the system of equations, the better will be the approximation at any given energy.

This expectation can be supported by a formal proof based on a variational principle (Spruch, 1967). However, the effect of increasing the number of channels taken into account does not lead to a uniform convergence of the cross section versus energy curve towards the exact one. Rather, the contrary can happen as follows from Figure 5.1. Here the results of the cross section calculation for the excitation of the 2s state in hydrogen are shown (Burke et al., 1963, 1967a, 1967b). It is seen that the two curves representing the results of calculations with three and six states taken into account differ strongly near the n = 3 thresholds.

It was shown in Section 4.6 that the closed channels produce at large distances a polarization potential $\sim r^{-4}$. Sometimes it is enough to take into account only one or two closed channel states

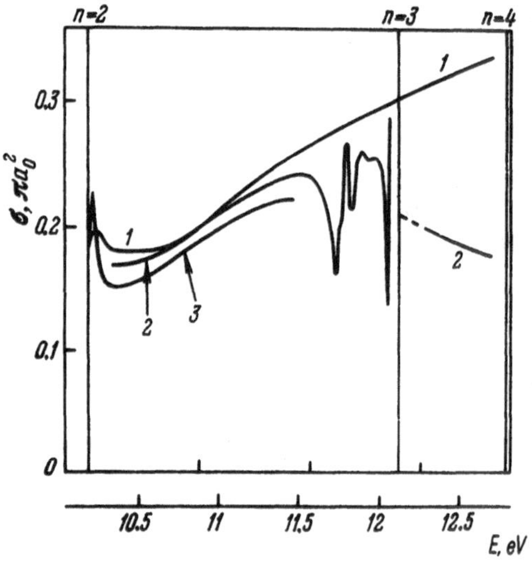

Fig. 5.1 2s - state excitation cross section in H:
1. 3-state close coupling calculation;
2. 6-state close coupling calculation;
3. 3-state close coupling + correlation terms calculation (Burke et al., 1963, 1967a, b).

to reproduce almost all the effect. For instance, the inclusion of only the first excited state for alkali metal atoms gives more than 90° of the polarizability.

An alternative way of treating the polarization of an atom by incident electron was suggested first by Damburg and Karule (1967). They proposed that so-called pseudo-states, which are target wave functions perturbed by an external uniform field, should be included in (1.1). This approach was used later by a number of authors.

In an attempt to diminish the error due to replacement of an infinite system of equations by a finite one, it was proposed to add a number of correlation terms Φ which are an orthonormal set of two-electron functions constructed from quadratically integrable one-electron functions, properly symmetrized. The typical form of Φ is

$$\Phi = r_1^{\alpha} r_2^{\beta} e^{-\kappa(r_1 + r_2)} |\mathbf{r}_1 - \mathbf{r}_2|^{\gamma}. \tag{1.2}$$

The parameters α, β, γ, κ are defined from the condition that the expression $\int \Psi^*(H - E)\Psi d\mathbf{r}_1 d\mathbf{r}_2$ should be stationary against small variations α, β, γ, κ and from the condition that Φ are orthogonal to the sum (1.1).

The effect of inclusion of such correlation terms is shown in Figure 5.1.

2. Matrix Variational Method

Instead of solving the coupled equations for the functions F_n entering in (1.1), one can introduce some trial functions depending on a set of parameters and determine them from the condition that the expression $\int \Psi_t^*(H - E)\Psi_t d\mathbf{r}_1 d\mathbf{r}_2$ should be stationary. Some correlation terms can be included as well. This is the idea of the matrix variational method (Nesbet, 1980). More explicitly a set of trial partial waves f_n^t is introduced, being regular at the origin and having the asymptotic form

$$f_n^t \sim (k_i^{-\frac{1}{2}}[\sin(k_i r - \frac{\ell_i \pi}{2})\delta_{ij} + K_{ij}^t \cos(k_j r - \frac{\ell_i \pi}{2})]. \tag{1.3}$$

The numbers K_{ij}^t form a trial K^t-matrix. Multiplying the partial wave by an angular part $Y_{\ell m}(\theta,\phi)$ and by $1/r$, a channel orbital function is formed which will be denoted by F_n^t. In addition, a number of quadratically integrable functions ϕ_s are introduced which are orthogonal to each other and to the partial waves f_n^t. From the functions ϕ_s, two-electron correlation functions Φ_s are formed. The trial total wave function is then constructed as

$$\Psi_t = \sum_{n=1}^{N} [\psi_n(\mathbf{r}_1)F_n^t(\mathbf{r}_2) \pm \psi_n(\mathbf{r}_2)F_n^t(\mathbf{r}_1)] + \sum_i c_i \Phi_i \tag{1.4}$$

and the expression

$$J = \int \Psi_t^* (H - E) \Psi_t dr_1 dr_2 \qquad (1.5)$$

is formed. This expression depends upon the parameters entering in Ψ and they are determined by stationary conditions.

According to the Kohn variational principle the K-matrix is given by the relation

$$K = St\{K^t - 2J\} \qquad (1.6)$$

where the symbol St denotes the stationary value.

Like the potential scattering case the matrix K can become infinite at certain energies. This singularity has no relation to real physical resonances. Near the energy at which K becomes singular one can use, instead of the K-matrix, its inverse for which

$$K^{-1} = St\{(K^t)^{-1} + 2J\}. \qquad (1.7)$$

This procedure is not completely satisfactory because it can introduce a discontinuity. More sophisticated procedures are considered by Nesbet in his book on variational methods (1980), together with many technical details including the search procedure of resonances.

3. R-Matrix Method

The essential idea of the method is that configuration space describing the scattered electron and the target atom is divided into two regions. The internal region is defined by a sphere of radius r_0, such that the charge distribution of the target states of interest are just contained within the sphere.

In the internal region where the scattered electron has penetrated the charge distribution of the target, the interaction is strong. Here electron exchange and correlation effects are important. However, in the external region exchange between the scattered electron and the target can be neglected, and the collision is dominated by the long range interactions only. Instead of matching directly the solutions of the Schrödinger equation in these two regions, a boundary condition on the surface of the sphere is used to determine the external solution (Burke and Robb, 1975)

$$f_p(r_0) = \sum_q R_{pq} r_0 \left(\frac{df_q}{dr}\right)_{r_0}. \qquad (1.8)$$

Here R_{pq} is the R-matrix. By looking for a solution with an asymptotic form like (1.3), one can express the K-matrix in terms of the R-matrix. Now, this R-matrix in turn should be determined by the solution in the internal region. A set of wave functions are introduced which imitate the target states together with an extra electron, and thus the R-matrix can be found. In fact it can be done in different ways. The most explicit description of the methods for the construction of the internal region wave function and R-matrix calculations can be found in a number of papers by Burke et al. (e.g., Burke and Robb, 1975; Burke, 1977). For an interesting discussion of a variational approach to the R-matrix, see Nesbet (1980).

The extension of above described methods for complex atoms meets two complications. The first is the rather complicated structure of the symmetry-adapted coordinate wave function of a system "electron + atom," expression (3.8) of Chapter 4 being the simplest example (except for the electron-hydrogen system). Sometimes it is possible to simplify the problem. For instance, when considering electron scattering by an alkali-metal atom, one can involve in the symmetrization procedure only the valence atomic electron and the scattered electron treating the closed atomic shells merely as an electric charge distribution. This approach is used in many practical calculations. In fact it can be a source of an error in the numerical value of the cross section which is not easy to estimate.

The second complication is the need to have accurate enough atomic wave functions. In many important cases there is a lack of them especially for excited states. This complication is even more serious than the first one. If the atomic state is described inadequately, all efforts of taking into account the effects of the interchannel interaction can become useless.

To illustrate the change in the calculated cross section due to the improvement of the target wave function we reproduce in Figure 5.2 the energy dependence of the e − O elastic scattering cross section calculated by the matrix variational method.

The three curves denoted by SC, CI, and BG correspond to the following approximations in the target wave function:

SC - the single configuration approximation which neglects the atomic polarizability.

CI - the configuration interaction approximation, which includes in the target-state variational basis the configurations $2s^2 2p^4$, $2s2p^5$ and $2p^6$. The inclusion of the $2s2p^5$ configuration contributes to the dipole polarizability, and the inclusion of the $2p^6$ configuration displaces the ^1S threshold relative to other atomic states. Both SC and CI calculations were carried out by Thomas et al. (1974).

BG - the Bethe-Goldstone approximation which takes into account the electron pair correlations and contains a more extended and accurate

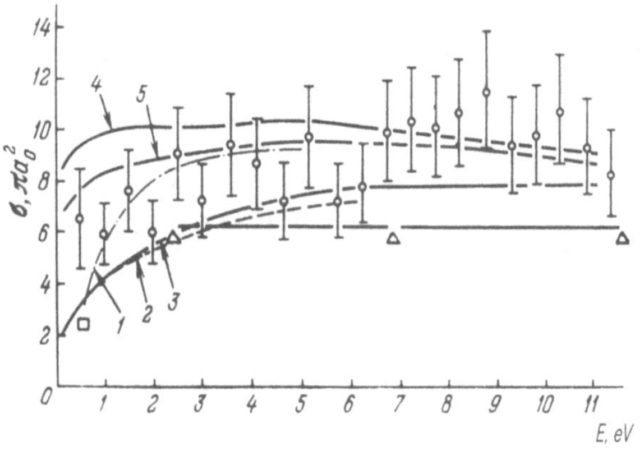

Fig. 5.2 e - O scattering. Theory:
1. Close coupling (Henry et al., 1969);
2. Polarized orbitals (Henry, 1967);
3. BG (Thomas and Nesbet, 1975);
4. SC (Thomas et al., 1974);
5. CI (Thomas et al., 1974)
(the meaning of BG, SC, and CI is explained in the text).
Experiment: ϕ – Sunshine et al. (1967); Δ – Neynaber et al. (1961); □ – Lin and Kivel (1959).

description of polarization. The calculations in the BG approximation were carried out by Thomas and Nesbet (1975). The experimental points including 20% error bars (Sunshine et al., 1967) are also shown in Figure 5.2.

5.2 THE BORN APPROXIMATION AND ITS MODIFICATIONS

In the Born approximation all interactions are treated as small and a procedure similar to perturbation theory is used. It can be justified for high incident electron energies. But how high? For potential scattering the condition for the validity of the Born approximation has been indicated (see Section 1.9 of Chapter 1).

But in the many-channel problem the situation is more complicated and no simple criterion can be formulated. However, some conclusions can be drawn from an exactly solvable model which imitates the excitation of the 2p state in H (Damburg and Propin, 1972). It turns out that even at energies 100 times higher than the excitation energy, the cross section calculated in the Born approximation is still 14% above the exact value. The reason for this rather slow convergence is the long range interactions.

The expression for the excitation amplitude in the Born approximation can be obtained from the standard system of equations for a many-channel problem like (3.5) of Chapter 4 if we put on the right hand side

$$F_\alpha^{S_t} = e^{i\mathbf{k}_0\cdot\mathbf{r}}\delta_{\alpha\alpha_0}.$$

This substitution corresponds just to the basic idea of the Born approximation: a plane wave represents the motion of a particle in the absence of any interaction.

Using the well-known technique a solution $F_\alpha^{(1)}$ of the equations satisfying the standard asymptotic conditions (3.7) of Chapter 4 can be found which is the first approximation to the exact solution. For the scattering amplitude we get:

$$A_{\alpha\alpha_0} = -\frac{1}{4\pi}\int e^{i\mathbf{q}\cdot\mathbf{r}}V_{\alpha\alpha_0}d\mathbf{r} - \frac{1}{4\pi}\int e^{i\mathbf{k}_\alpha\cdot\mathbf{r}}d\mathbf{r}\int W_{\alpha\alpha_0}(\mathbf{r}\cdot\mathbf{r}')e^{i\mathbf{k}_0\cdot\mathbf{r}'}d\mathbf{r}'. \quad (2.1)$$

Here $\mathbf{q} = \mathbf{k}_0 - \mathbf{k}_\alpha$ is the momentum transfer. The first term on the right hand side is called the direct amplitude and the second is called the exchange amplitude.

Let us consider first the direct amplitude

$$A'_{\alpha\alpha_0} = -\frac{1}{4\pi}\int e^{i\mathbf{q}\cdot\mathbf{r}}V_{\alpha\alpha_0}d\mathbf{r}. \quad (2.2)$$

According to (3.6) of Chapter 4, for the hydrogen atom

$$V_{\alpha\alpha_0} = 2\int\frac{\psi_\alpha^*(\mathbf{r}')\psi_{\alpha_0}(\mathbf{r}')}{|\mathbf{r}-\mathbf{r}'|}d\mathbf{r}' - \frac{2}{r}\delta_{\alpha\alpha_0}.$$

For a complex atom

$$V_{\alpha\alpha_0} = 2\sum_i\int\frac{\psi_\alpha^*\psi_{\alpha_0}}{|\mathbf{r}-\mathbf{r}_i|}d\mathbf{r}_1\ldots d\mathbf{r}_n - \frac{2Z}{r}\delta_{\alpha\alpha_0}.$$

Using the relation

$$\int\frac{e^{i\mathbf{q}\cdot\mathbf{r}}}{|\mathbf{r}-\mathbf{r}'|}d\mathbf{r} = -\frac{4\pi}{q^2}e^{i\mathbf{q}\cdot\mathbf{r}'},$$

it is possible to rewrite $A'_{\alpha\alpha_0}$

$$A_{\alpha\alpha_0} = \frac{2}{q^2}\left[\sum_i\int e^{i\mathbf{q}\cdot\mathbf{r}_i}\psi_\alpha^*\psi_{\alpha_0}d\mathbf{r}_1\ldots d\mathbf{r}_n - Z\delta_{\alpha\alpha_0}\right]. \quad (2.3)$$

In the case of elastic scattering ($\alpha = \alpha_0$), (2.3) gives

$$A' = \frac{2}{q^2}[F(q) - Z] \quad (2.4)$$

where

$$F(q) = \sum_i \int |\psi_{\alpha 0}|^2 e^{i\mathbf{q}\cdot\mathbf{r}_i} d\mathbf{r}_1 \ldots d\mathbf{r}_n \tag{2.5}$$

is the atomic form factor. This expression is similar to (9.23) of Chapter 1. In the case of inelastic scattering

$$A_{\alpha\alpha^0} = \frac{2}{q^2} \sum_i \int e^{i\mathbf{q}\cdot\mathbf{r}_i} \psi_\alpha^* \psi_{\alpha^0} d\mathbf{r}_1 \ldots d\mathbf{r}_n . \tag{2.6}$$

The expression (2.6) can be transformed further if we realize that

$$\mathcal{A} = \frac{2}{q^2} \sum_i e^{i\mathbf{q}\cdot\mathbf{r}_i}$$

is in fact the sum of scattering amplitudes from point charges located at the positions $\mathbf{r}_1 \ldots \mathbf{r}_n$. Then

$$A_{\alpha\alpha^0} = \int \psi_\alpha^* \mathcal{A} \psi_{\alpha^0} d\mathbf{r}_1 \ldots d\mathbf{r}_n . \tag{2.7}$$

The expression of this type is called the adiabatic approximation.

The presence of the oscillating factor exp $i\mathbf{q}\cdot\mathbf{r}_i$ in the integrand leads to a decreasing amplitude at large q. For example, the amplitude of n = 2 level excitation in hydrogen decreases like q^{-4} when q >> 1. It follows that the main contribution comes from the momentum transfer not exceeding several atomic units.

We turn now to the exchange amplitude:

$$A''_{\alpha\alpha^0} = -\frac{1}{4\pi} \int e^{-i\mathbf{k}_\alpha\cdot\mathbf{r}} d\mathbf{r} \int W_{\alpha\alpha^0}(\mathbf{r},\mathbf{r}') e^{i\mathbf{k}_0\cdot\mathbf{r}'} d\mathbf{r}' . \tag{2.8}$$

Let us consider the case of a hydrogen atom as a target. Taking into account the expression for $W_{\alpha\alpha^0}$ (3.6) of Chapter 4, we have

$$A''_{\alpha\alpha^0} = (-1)^{S_t} \frac{1}{2\pi} \int \psi_\alpha^*(\mathbf{r}) \psi_{\alpha^0}(\mathbf{r}') e^{i(\mathbf{k}_0\cdot\mathbf{r}' - \mathbf{k}_\alpha\cdot\mathbf{r})} \left[\frac{1}{|\mathbf{r}-\mathbf{r}'|} + E_\alpha - \frac{k_0^2}{2} \right] d\mathbf{r} d\mathbf{r}' . \tag{2.9}$$

Considering the asymptotic behavior of (2.9) when $k_0 \to \infty$, we can see that there are several terms decreasing like k^{-n} with different n. Following Ochkur's reasoning (Ochkur, 1963), we should retain only the term with the smallest n and omit all others. This is because the Born approximation itself is an asymptotic expression. If we retain the fast decreasing terms this would be inconsistent with the accuracy of the method.

To separate the leading term it is advantageous to use the Fourier transform of the atomic wave function (or the momentum representation). The result is

$$A''_{\alpha\alpha^0} = (-1)^{S_t} \frac{2}{k_0^2} \int e^{i\mathbf{q}\cdot\mathbf{r}} \psi_\alpha^* \psi_{\alpha^0} d\mathbf{r} . \tag{2.10}$$

This expression is called the Ochkur approximation. We will write down also the amplitude for the singlet-triplet transition in helium

$$A''_{\gamma\beta 0} = -\frac{\sqrt{3}}{k_o^2}\int \psi_\gamma^*(;\mathbf{r}_1\mathbf{r}_2)\psi_{\beta 0}(\mathbf{r}_1;\mathbf{r}_2)e^{i\mathbf{q}\cdot\mathbf{r}_1}d\mathbf{r}_1 d\mathbf{r}_2. \qquad (2.11)$$

It follows that the exchange amplitudes are proportional to $1/k_o^2$ while the direct amplitudes are proportional to $1/q^2$. So in the high-energy region we can neglect the exchange amplitudes. The corresponding cross section will be equal to

$$\frac{d\sigma_{\alpha\alpha 0}}{d\Omega} = \frac{k_\alpha}{k_o}\frac{4}{q^4}\left|\sum_i \int e^{i\mathbf{q}\cdot\mathbf{r}_i}\psi_\alpha^*\psi_{\alpha 0}d\mathbf{r}_1\ldots d\mathbf{r}_n\right|^2.$$

It is customary to use the generalized oscillator strength which is defined as

$$f_{\alpha\alpha 0} = \frac{2\Delta E}{q^2}\left|\sum_i \int e^{i\mathbf{q}\cdot\mathbf{r}_i}\psi_\alpha^*\psi_{\alpha 0}d\mathbf{r}_1\ldots d\mathbf{r}_n\right|^2. \qquad (2.12)$$

Here

$$\Delta E = \frac{1}{2}(k_o^2 - k_\alpha^2)$$

is the energy transfer. In terms of the generalized oscillator strength

$$\frac{d\sigma_{\alpha\alpha 0}}{d\Omega} = 2\,\frac{k_\alpha}{k_o}\frac{1}{q^2\Delta E}f_{\alpha\alpha 0}(q). \qquad (2.13)$$

In the case when the final state belongs to a continuum, that is, in the case of ionization, a continuum spectrum wave function $\psi(\mathbf{k}',\mathbf{r})$ $(\frac{k'^2}{2} = \varepsilon)$ will appear instead of ψ_α. Then instead of the oscillator strength, the spectral density $df/d\varepsilon$ is used.

The total cross section is obtained by integrating (2.13) over the angular variables. It is advantageous to integrate over the momentum transfer q instead of the angular variables. Using the relation

$$q^2 = k_\alpha^2 + k_o^2 - 2k_\alpha k_o\cos\theta,$$

we can write

$$k_\alpha k_o\sin\theta d\theta = q\,dq.$$

Then

$$d\sigma_{\alpha\alpha 0} = \frac{4\pi}{k_o^2}\frac{1}{q\Delta E}f_{\alpha\alpha 0}dq, \qquad (2.14)$$

and

$$\sigma_{\alpha\alpha^0} = \frac{4\pi}{k_o^2} \int_{q_{min}}^{q_{max}} f_{\alpha\alpha^0}(q) \frac{dq}{q} \tag{2.15}$$

where

$$q_{min} = k_o - k_\alpha, \quad q_{max} = k_o + k_\alpha.$$

It is easy to observe the difference between optically allowed and optically forbidden transitions using expression (2.15).

For optically allowed transitions $f_{\alpha\alpha^0}(0) \neq 0$, and the main contribution to the integral comes from the region near the lower limit. For forbidden transitions $f_{\alpha\alpha^0}(0) = 0$, and the main contribution comes from the region $q \sim 1$.

For optically allowed transitions the expression (2.15) can be simplified. Let us write $f_{\alpha\alpha^0}(0)$ instead of $f_{\alpha\alpha^0}(q)$ and change the upper limit to q' where the value of the integral is unchanged. Then

$$\sigma_{\alpha\alpha^0} = \frac{4\pi}{k_o^2 E} f_{\alpha\alpha^0}(0) \ln \frac{q'}{q_{min}} .$$

For high enough energy one can put $q' \simeq \text{const} \cdot k_o$. Then

$$\frac{q'}{q_{min}} \simeq \beta \frac{E_o}{\Delta E}$$

and

$$\sigma_{\alpha\alpha^0} = \frac{2\pi}{E_o \Delta E} f_{\alpha\alpha^0}(0) \ln \frac{\beta E_o}{\Delta E} . \tag{2.16}$$

This is the Bethe approximation. The numerical value of β depends upon details of the atomic structure and of the transition in question.

Various aspects of the Bethe approximation are considered by Inokuti (1971) in a review article.

Note that a scaling relation can be deduced from (2.16):

$$\sigma E_o^2 = \mathcal{F}\left(\frac{E_o}{\Delta E}\right) \tag{2.17}$$

where $\mathcal{F}$ is a universal function.

If an electron is scattered by ion of a charge Z then it seems natural to use in the integrand (2.1) Coulomb wave functions instead of plane waves. This kind of approximation is called the Coulomb-Born approximation. An interesting property of this approximation is the scaling relation which emerges when $Z \gg 1$ for hydrogen-like ions

$$\sigma Z^4 = f(k/Z). \tag{2.18}$$

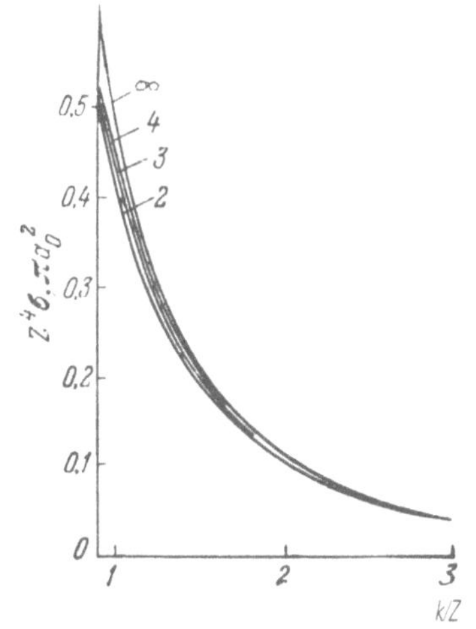

Fig. 5.3 2s state excitation cross section for hydrogen-like ions with various Z. $Z^4\sigma$ is plotted against k/Z (Gailitis, 1963).

Figure 5.3 shows the cross sections for the 2s-state excitation of a hydrogen-like ion for different values of Z. The curves are very close to each other using reduced coordinates $Z^4\sigma$ versus k/Z (Gailitis, 1963) even for quite small Z.

A remarkable property of the Born approximation for the direct amplitude (2.6) and the Ochkur approximation for the exchange amplitude (2.10) is the ability to reproduce the energy dependence of cross sections averaged over the resonances within a factor 1.5 - 2 even near excitation thresholds.

Let us consider for example the excitation of the hydrogen atom. The cross section which follows from (2.6) and (2.10) is

$$\sigma_{\alpha\alpha^0} = \frac{4\pi}{k_0^2 \Delta E} \int_{q_{min}}^{q_{max}} f_{\alpha\alpha^0}(q) \left[\frac{1}{q^2} \pm \frac{1}{k_0^2} \right]^2 q^3 dq. \qquad (2.19)$$

The ± sign refers to the two states of total spin: S = 0 or 1. Averaging over the spin states we get

$$\bar{\sigma}_{\alpha\alpha^0} = \frac{4\pi}{k_0^2 \Delta E} \int_{q_{min}}^{q_{max}} f_{\alpha\alpha^0} \left(\frac{1}{q^4} + \frac{1}{k_0^4} - \frac{1}{q^2 k_0^2} \right) q^3 dq. \qquad (2.20)$$

Figure 5.4 shows the cross section for excitation of the 2p-state calculated according to (2.20) (curve 2) the cross section calculated without the exchange terms (curve 1), and the experimental results (curve 3).

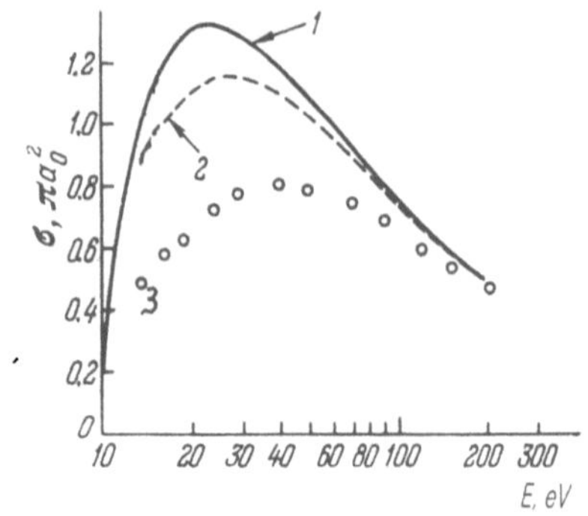

Fig. 5.4 2p-state excitation cross section for H:
1. Born approximation without exchange;
2. Cross section according to (2.20);
3. Experimental data.

It follows that the Born and Ochkur approximations are a good starting point for various interpolation and semi-empirical expressions intended to bring the calculated cross section averaged over the resonances closer to experiment.

Let us now return to the beginning of this section and discuss how to improve the Born approximation using again the system of equations (3.5) of Chapter 4.

If instead of a plane wave we substitute on the right side the first approximation $F_\alpha^{(1)}$ and solve the equations again, we get the second approximation $F_\alpha^{(2)}$. From this solution the second Born approximation can be derived for the amplitude which is again a sum of direct and exchange terms.

Like the second order correction to the energy eigenvalue in the case of a bound state perturbation, the second Born approximation can be interpreted as a correction due to virtual transitions through intermediate states.

It is quite obvious to expect the most significant improvement in the cases when the first Born approximation leads to small cross sections.

It follows from the expression (2.6) for the direct amplitude that it becomes small at large scattering angles and high energy when $q \gg 1$. Then one can expect an important contribution of virtual transitions through intermediate states in the differential cross section at large angles. As for the total cross section, a significant contribution of transitions through intermediate states can be expected when the first Born approximation is lower than the experimental value.

It can be seen in Figure 5.4 that for the 2p-state excitation the calculated cross section is higher than the experimental one. However, the situation is not always the same. For instance, the cross section for 3^1D state excitation in He calculated in the first Born approximation is lower than experiment.

The large scale calculation of second Born approximation amplitudes is rather tedious. However, Ochkur's reasoning can be repeated and leads to the conclusion that one should not even try to perform such a calculation. Instead, the leading term in the high-energy asymptotic form should be separated and only it should be used.

This procedure was carried out by Ochkur and comparatively simple expressions for the direct and exchange amplitudes were obtained (Burkova and Ochkur, 1979).

As an example the excitation cross section of the 3^1D state in He is shown on Figure 5.5.

5.3 THE GLAUBER APPROXIMATION

The Glauber approximation for a system of fixed centers was considered in Chapter 1.

To understand how this approximation can be applied to electron-atom scattering, let us first consider elastic scattering.

Suppose that the velocity of the incident electron is much higher than the velocity of the atomic electrons. Then during the passage through the atom its electrons can be considered at rest in some fixed positions. In each collision the positions of the electrons during the collision will be different, and the scattering amplitude should be averaged over all possible positions of the

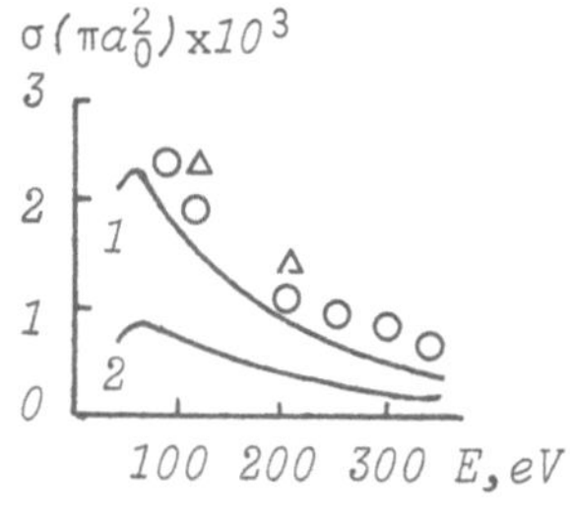

Fig. 5.5 Excitation cross section for 3^1D state in He:
1. Asymptotic second Born approximation (Burkova and Ochkur, 1979);
2. First Born approximation;
Δ, O – experimental data.

atomic electrons. This averaging is achieved by first multiplying the Glauber amplitude $\mathcal{A}$ for fixed electrons by the probability of a given distribution of electrons and then summing over all possible positions. That is,

$$A = \int |\psi(\mathbf{r}_1 \ldots \mathbf{r}_n)|^2 \mathcal{A} \, d\mathbf{r}_1 \ldots d\mathbf{r}_n . \tag{2.21}$$

For inelastic scattering the amplitude $A_{\alpha\alpha^0}$ is equal to

$$A_{\alpha\alpha^0} = \int \psi_\alpha^* \mathcal{A} \psi_{\alpha^0} d\mathbf{r}_1 \ldots d\mathbf{r}_n \tag{2.22}$$

(Glauber, 1959).

Figure 5.6 shows the results of the 2p-state excitation cross section in H calculated in the Glauber approximation (Tai et al., 1970), the first Born approximation and experimental data being shown for comparison.

It can be seen that the Glauber approximation is valid at a much lower energy than one can expect from the nature of the approximation. The Glauber approximation in a similar way to the second Born approximation improves the differential cross section calculated in the first Born approximation at large angles for inelastic scattering.

However, for elastic scattering the main correction occurs at small angles (Gerjuoy, 1971). The reason is that, unlike the first Born approximation in the Glauber approximation, the interaction with the atomic nucleus is taken into account even for inelastic scattering. This interaction is most important at large scattering

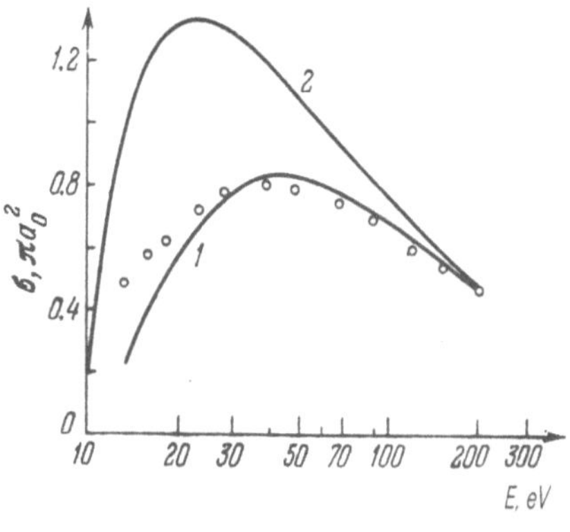

Fig. 5.6 2p-state excitation cross section for H:
1. Glauber approximation;
2. Born approximation;
0 – Experimental data (Tai et al., 1970).

angles. However, in the case of elastic scattering both approximations take into account the interaction with the atomic nucleus.

5.4 CLASSICAL MECHANICS (BINARY APPROXIMATION)

Classical mechanics can be applied to calculate the cross section for energy transfer to the atomic electrons. Supposing that the energy transfer leads ultimately to excitation or to ionization of the atom, we will then identify the cross section for a given energy transfer as the cross section of excitation or ionization.

Strictly speaking, even in the simplest case involving an electron collision with a hydrogen atom, the energy transfer calculation is a three-body problem. However, for the sake of estimation the problem can be simplified by neglecting the interaction of the incident electron with the nucleus. Then the problem reduces to a two-body collision of an incident electron with an atomic electron. This approach is called the binary approximation. The calculation of a given energy transfer cross section in this approximation is very simple. No more than the Rutherford formula is needed. The cross section for the energy transfer in an interval ε, $\varepsilon + d\varepsilon$ from an incident electron having an energy E to an atomic electron (which is initially at rest), is equal to

$$\frac{d\sigma}{d\varepsilon} = \frac{\pi}{E}\frac{1}{\varepsilon^2} \tag{4.1}$$

(we remember that in atomic units the electric charge is taken as unity). It follows that from (4.1) the ionization cross section

$$\sigma = n\,\frac{\pi}{E}\int_I^E \frac{d\varepsilon}{\varepsilon^2} = n\,\frac{\pi}{E}\left(\frac{1}{I} - \frac{1}{E}\right) \tag{4.2}$$

where I is the ionization potential and n is the number of electrons in the atomic shell with ionization potential I.

This expression can be rewritten in the form

$$\frac{I^2\sigma}{n} = F\left(\frac{E}{I}\right) \tag{4.3}$$

where F is a universal function which according to (4.2) is equal to

$$F(x) = \frac{\pi}{x}\left(1 - \frac{1}{x}\right).$$

This is the scaling relation for ionization which holds when the above binary approximation is valid.

However, to consider an atomic electron as being at rest is rather an oversimplification. A more realistic approximation will be to assume a certain average velocity of atomic electrons or even a distribution of electron velocities. We reproduce here the result obtained by Stabler (1964) using more realistic assumptions about the velocity of the atomic electrons.

Let the velocity of the atomic electron be v_1 and that of the incident electron be v_2. Then

$$\frac{d\sigma}{d\varepsilon} = \frac{\pi}{v_1}\left(\frac{2B}{E_1E_2}\right)^{\frac{1}{2}}\left(1+\frac{4}{3}\frac{B}{\varepsilon}\right)\frac{1}{\varepsilon^2} \tag{4.4}$$

where

$$B = \min\{E_1, E_2, E_1', E_2'\}$$

is the lowest of the energy levels before (E_1, E_2) and after (E_1', E_2') the collision.

The most natural value of v_1 will be

$$v_1 = \sqrt{2I} \equiv v_o \tag{4.5}$$

which follows from the virial theorem.

The ionization cross section is still given by (4.3). However, the function F is more complicated

$$F(x) = \frac{8}{3x}(1-x)^{3/2} \quad 1 \leqslant x \leqslant 2,$$

$$F(x) = \frac{4}{x}\left(\frac{5}{3} - \frac{1}{x-1}\right) \quad x > 2.$$

The cross section for an inelastic collision corresponding to the excitation in an assembly of energy levels between I_n and I_{n+1} is equal to

$$\sigma = \int_{I_n - I_o}^{I_{n+1} - I_o} \frac{d\sigma}{d\varepsilon}\, d\varepsilon \quad E_1 \geqslant I_{n+1} - I_o,$$

$$\sigma = \int_{I_n - I_o}^{I_n} \frac{d\sigma}{d\varepsilon}\, d\varepsilon \quad I_n - I_o \leqslant E_1 \leqslant I_{n+1} - I_o. \tag{4.6}$$

The cross section for exchange scattering is defined as

$$\sigma_{ex} = \int_{E_1 + I_n}^{E_1 + I_{n+1}} \frac{d\sigma}{d\varepsilon}\, d\varepsilon. \tag{4.7}$$

Note that according to (4.2) the asymptotic form of σ at large E is

$$\sigma \sim \text{const}/E, \tag{4.8}$$

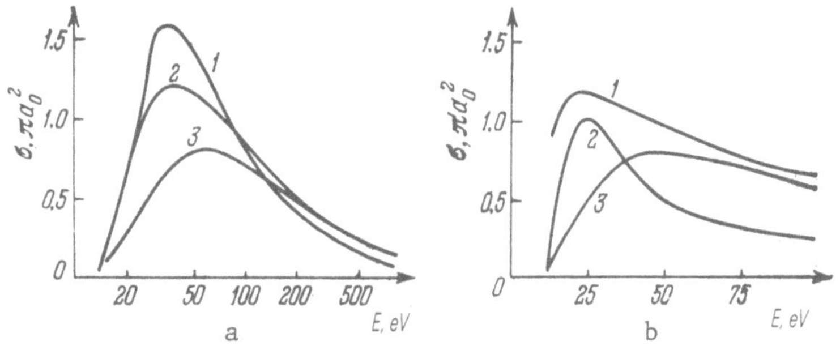

Fig. 5.7 Total 2s + 2p excitation cross section for H (a) and ionization cross section (b):
1. Born approximation;
2. Classical binary calculation;
3. Experimental data.

while in the Bethe approximation (2.16) it is

$$\sigma \sim a/E + B \ln E. \tag{4.9}$$

The logarithmic term cannot be reproduced by classical mechanics but it can be obtained by means of a more refined semiclassical treatment (Beigman, Vainstein and Sobelman, 1969).

Figure 5.7a shows the total cross section for excitation of the 2s + 2p states in hydrogen based on (4.4) and Figure 5.7b shows the ionization cross section. For comparison the results of the Born approximation and experimental data are also shown.

6
Elastic and Inelastic Scattering of Electrons by Atoms and Positive Ions

6.1 HYDROGEN

A. Energy of the Incident Electron Below the First Excitation Threshold

In this region only elastic scattering is possible. There are a large number of phase shifts and scattering length calculations. One of the most accurate is based on the variational method (Schwartz, 1961). The wave function of the "electron + atom" system is represented in the form

$$\Psi = \psi_o(\mathbf{r}_1)F_o(\mathbf{r}_2) \pm \psi_o(\mathbf{r}_2)F_o(\mathbf{r}_1) + \sum_i c_i\Phi_i.$$

Here ψ_o is the ground state atomic wave function, F_o is the continuum state wave function having the standard asymptotic form, and Φ_i is a set of quadratic-integrable correlation functions. The variable parameters entering F_o and Φ_i are determined by the condition of stationarity of the functional

$$\int \Psi^* (H - E)\Psi d\mathbf{r}_1 d\mathbf{r}_2.$$

The following numbers were obtained for the scattering length: triplet state $a^{(1)} = 1.768$, singlet state $a^{(0)} = 5.961$.

Note that both scattering lengths are positive. The calculated values of the S phases at zero energy are equal to π for singlet and triplet states.

At first sight this contradicts Levinson's theorem because there is a bound state of the hydrogen negative ion in the singlet state

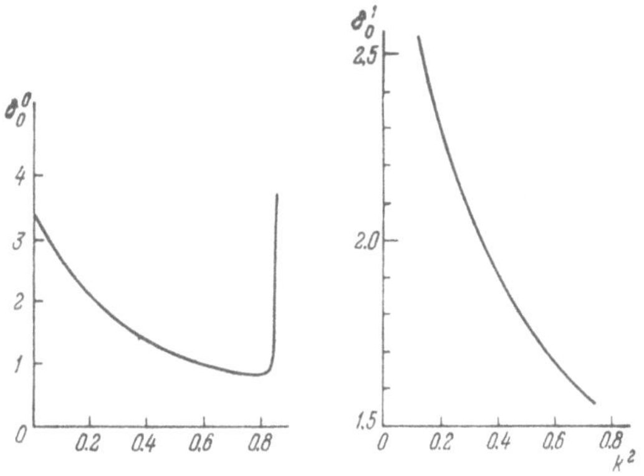

Fig. 6.1 e - H scattering: singlet and triplet phases, $\ell = 0$ (Gailitis, 1965).

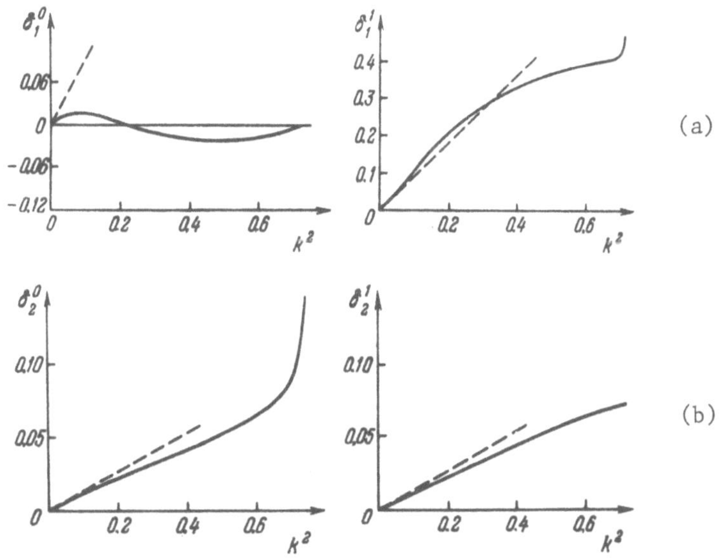

Fig. 6.2 e - H scattering: singlet and triplet phases, $\ell = 1$ (a), $\ell = 2$ (b) (Gailitis, 1965).
Dashed curve calculated according to (5.13) of Chapter 1.

but no stable negative ion in the triplet state. It seems that in the case of several identical particles entering in the system, Levinson's theorem should count a bound state forbidden by the Pauli principle as a real one. Another approach was used by Gailitis (1965). The functions F and coefficients c_i were determined from a system of coupled integrodifferential and algebraic equations, and the parameters entering in Φ_i were determined by a variational method.

In Figures 6.1 and 6.2 the singlet and triplet phases for $\ell = 0, 1, 2$ are shown according to Gailitis. Actually the lowest value of k^2 for which the phases have been calculated was 0.01.

Phases calculated according to the expression

$$\delta_\ell = \frac{\pi\alpha k^2}{8(\ell + 3/2)(\ell + 1/2)(\ell - 1/2)} \tag{1.1}$$

[see equation (5.13) of Chapter 1] are shown also. It can be seen that for $\ell = 2$ this expression is a good approximation over a wide energy range.

More recent calculations of phases (also starting at the point $k^2 = 0.01$) have been reviewed by Callaway (1978).

Angular Distribution and Total Elastic Cross Section at Low Energy. Using calculated values of phases one can determine the amplitudes for singlet and triplet scattering:
tudes for singlet and triplet scattering

$$A^{S_t} = \frac{1}{2ik} \sum_{\ell = 0}^{\infty} (2\ell + 1)[\exp 2i\delta_\ell^{S_t} - 1]P_\ell(\cos\theta)$$

($S_t = 0$ or 1), the corresponding differential cross sections

$$\frac{d\sigma^{S_t}}{d\Omega} = |A^{S_t}|^2,$$

and the spin-averaged differential cross section

$$\frac{d\sigma}{d\Omega} = \frac{3}{4}|A^1|^2 + \frac{1}{4}|A^0|^2.$$

Following Gailitis (1965) and Thompson (1966), let us divide the sum over the angular momentum quantum number ℓ into 2 parts:

$$\frac{1}{2ik}\sum_{\ell=0}^{\infty}(2\ell + 1)(e^{2i\delta_\ell} - 1)P_\ell(\cos\theta)$$

$$= \frac{1}{2ik}\sum_{\ell=0}^{2}(2\ell + 1)(e^{2i\delta_\ell} - 1)P_\ell(\cos\theta) + \frac{1}{2ik}\sum_{\ell=3}^{\infty}(2\ell + 1)(e^{2i\delta_\ell} - 1)P_\ell(\cos\theta).$$

In the first sum the phases computed by variational methods or numerical integration of coupled equations should be used. In the second sum we will use the approximate expression (1.1). With the help of equations (5.12) and (5.13) of Chapter 1 we can evaluate the sum over ℓ in the second sum {adding and subtracting $\frac{1}{2ik}\sum_{\ell=1}^{2} P_\ell \times \cos\theta/[(2\ell+3)(2\ell+1)]$} and obtain finally the following expression

$$\frac{1}{2ik}\sum_{\ell=0}^{\infty}(2\ell+1)(e^{2i\delta_\ell}-1)P_\ell(\cos\theta)$$
$$=\frac{1}{2ik}\sum_{\ell=0}^{2}(2\ell+1)(e^{2i\delta_\ell}-1)P_\ell(\cos\theta)$$
$$+\pi\alpha k[(\frac{1}{3}-\frac{1}{2}\sin\frac{\theta}{2})-\sum_{\ell=1}^{2}\frac{P_\ell(\cos\theta)}{(2\ell+1)(2\ell-1)}] \qquad (1.2)$$

which was used to compute the differential and total cross section.

In Figure 6.3 the differential cross section for several energies calculated by Gailitis is shown.

Note that the finite slope of the curves at $\theta = 0$ is due to the action of the long range tail $\sim r^{-4}$ in the potential (see Section 1.5). The sign of $\frac{d}{d\theta}(\frac{d\sigma}{d\Omega})_{\theta-0}$ is due to the negative sign in (5.16) of Chapter 1.

Figure 6.4 shows the absolute value of the measured differential cross section (Williams, 1974) at 3.4 eV together with the theoretical values calculated by various methods. Note a marked anisotropy even at such a low energy as (k = 1/2) which is due to the exchange interaction. It was found by John (1960) that for ℓ = 1 the exchange interaction significantly increases the value of the ℓ = 1 phase.

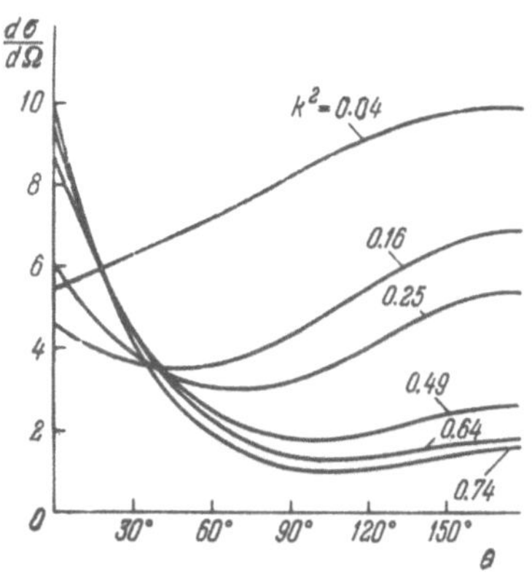

Fig. 6.3 Differential cross section at various energies (Gailitis, 1965).

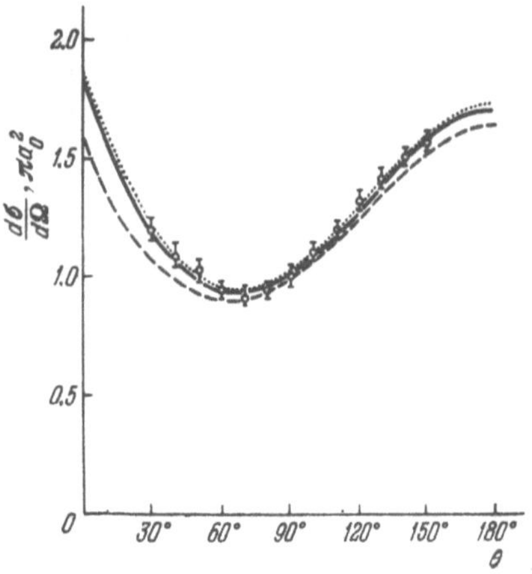

Fig. 6.4 e - H scattering: differential cross section at 3.4 eV. Dotted line, dashed line and full line - various versions of the theoretical calculation with s, p and d phases taken into account;
ϕ - experimental data (Williams, 1974).

Figure 6.5 shows the energy dependence of the total spin-averaged elastic scattering cross section calculated by Gailitis (1965). The energy interval covered in this calculation is chosen not to be too close to the resonance region. Here again we should note the effect of the long range tail $\sim r^{-4}$ in the potential. According to (5.7) of Chapter 1 the derivative $(\frac{d\sigma}{dk})_{k=0}$ is proportional to the product $a\alpha$. Because $a > 0$ the derivative $(\frac{d\sigma}{dk})_{k=0}$ will be positive. [If we plot σ against k^2 rather than k, then we get $(\frac{d\sigma}{dk^2})_{k=0} = \infty$.] Unfortunately this initial rise of the cross section and subsequent transition through a maximum is located in the region $k < 0.1$ which was not covered by any of the existing calculations.

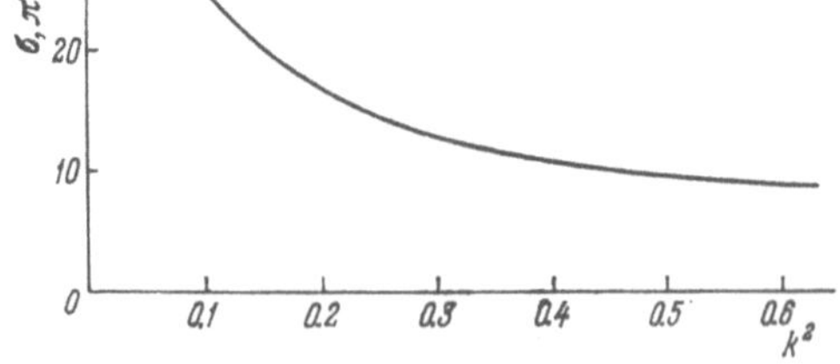

Fig. 6.5 Elastic scattering cross section for e - H (Gailitis, 1965).

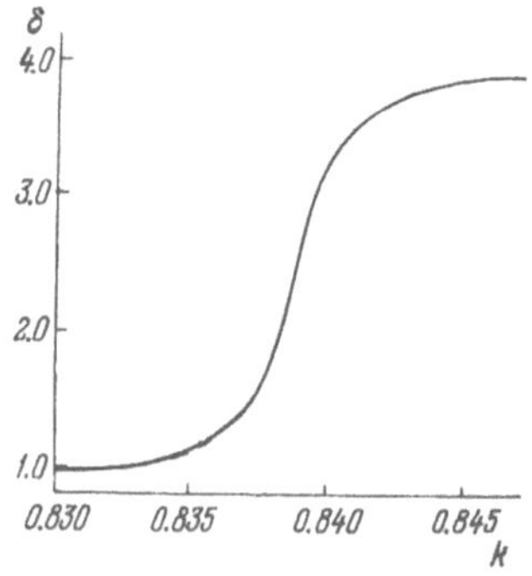

Fig. 6.6 e - H scattering, singlet S phase near the resonance.

Resonances. The specific feature of the hydrogen atom is the degeneracy of the excited states with respect to the angular momentum quantum number ℓ. This leads to a long range interaction of an electron with an excited atom with a tail $\sim r^{-2}$, which can bind an infinite number of quasistationary states below the n = 2 threshold (Gailitis and Damburg, 1963). There is a characteristic behavior of the phase near the energy of each quasistationary state, similar to that in potential scattering (Chapter 1, Section 1.4).

As an example in Figure 6.6 the energy dependence of the singlet S-wave phase is shown calculated by Nesbet and Lyons (1971). They used a variational method and a special resonance-search procedure.

Table 1 (taken from Nesbet, 1980) compares the results of the most accurate calculations for the three lowest resonances with experimental data. Figure 6.7 shows the corresponding change in the measured intensity of an electron beam scattered at 90° (Schulz, 1973).

Table 1. Variational Calculation of Resonances in e - H Elastic Scattering

State	Energy (eV)		Width (eV)	
	Theory	Experiment	Theory	Experiment
^1S	9.557	9.558 ± 0.01	0.04720	0.0430 ± 0.0060
$^3P^o$	9.738	9.738 ± 0.01	0.00585	0.0056 ± 0.0055
^1D	10.122	10.128 ± 0.01	0.00900	0.0073 ± 0.0020

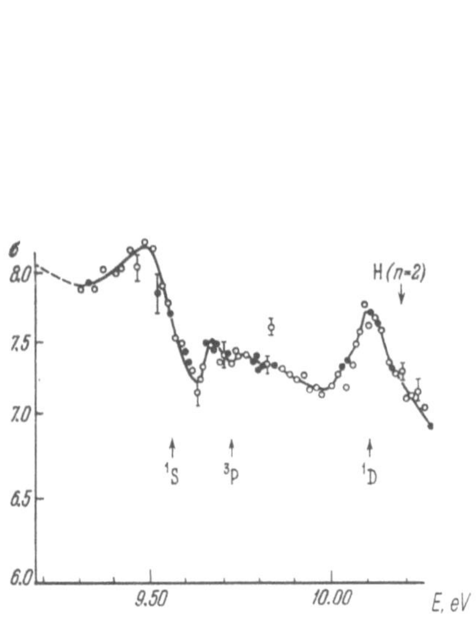

Fig. 6.7 e - H scattering, resonance structure of the scattered electron current intensity. The quantum numbers of the resonances and their positions are indicated by arrows (Schulz, 1973).

B. Energy Region between the n = 2 and n = 3 Excitation Thresholds

In this region elastic scattering and excitation of the 2s and 2p states are possible. Close coupling plus correlation term calculations were performed by Taylor and Burke (1967), Burke et al. (1967), and Macek and Burke (1967). Variational calculations have been recently reviewed by Callaway (1978). Just above the n = 2 threshold and near the n = 3 threshold the cross sections are affected by resonances.

First let us consider the resonance near the n = 2 threshold. It is located at the energy $E \approx 10.22$ eV and its width is ≈ 0.015 eV. This resonance is due to a quasistationary ^1P state of the negative hydrogen ion H^- temporarily formed during the collision. It produces a sharp structure in the excitation cross sections for the 2s and 2p states as shown in Figures 6.8 and 6.9.

The resonances below the n = 3 threshold for total angular momentum $\ell_t \leqslant 3$ are listed in Table 2 taken from the review paper by Callaway (1978).

The position E_R and the width Γ of the resonances are related to the sum of the eigenphases of the scattering matrix Callaway, 1978) by:

$$\Sigma\Delta_i = a + bE + \arctan\frac{\Gamma}{2(E_R - E)} \,. \tag{1.3}$$

The first two terms on the right side of (1.3), a + bE, represent the slowly changing background contribution while the third term, $\arctan \dfrac{\Gamma}{2(E_R-E)}$, is responsible for the rapid change in $\Sigma\Delta_i$. As

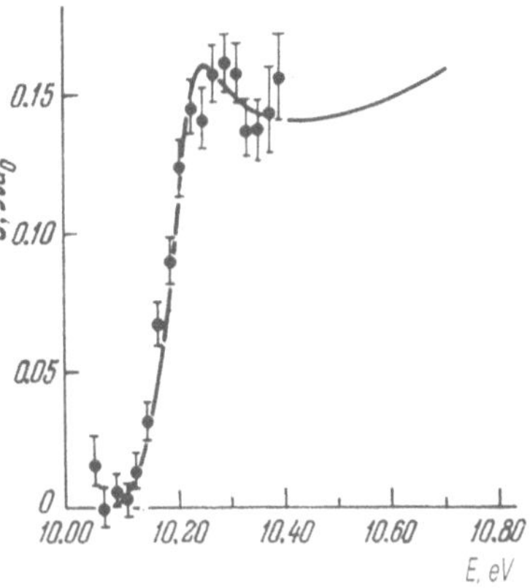

Fig. 6.8 2s-state excitation cross section for H:
Dots - experiment; curve-close coupling plus correlation term calculation averaged over an energy distribution of 0.08 eV (Elston et al., 1975).

an example the eigenphase sum for the ^1S state can be mentioned. It is equal to - 0.464 at $k^2 = 0.76$ and to + 1.695 at $k^2 = 0.78$.

The calculated excitation cross sections for the 2s and 2p-states together with the experimental measurements are shown in Figures 6.10 and 6.11 (Callaway, 1978; experimental points - Williams, 1976).

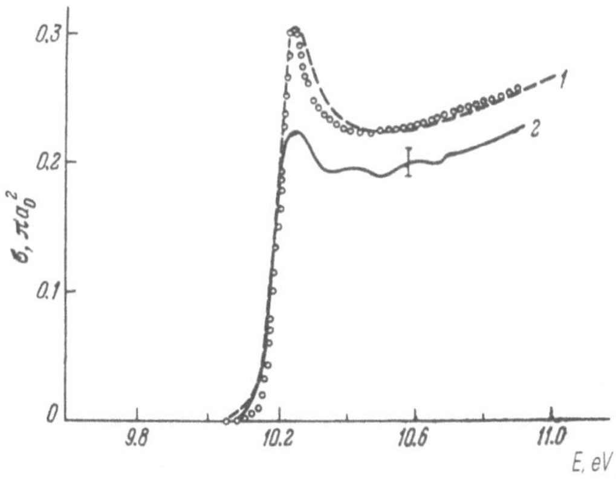

Fig. 6.9 2p-state excitation cross section for H:
1, 2 - experiment; dotted line - close coupling plus correlation term calculation averaged over an energy distribution of 0.07 eV (Williams and Willis, 1974).

Table 2. Resonances below the n = 3 Threshold

State	E_r(Ry)	$\Gamma_r(10^{-4}$Ry)
$^1S(1)$	0.8621	28.3
$^1S(2)$	0.88461	5.82
$^3S(1)$	0.88212	0.18
$^3S(2)$	0.88767	0.09
$^1P(1)$	0.87495	24.2
$^1P(2)$	0.88297	0.09
$^1P(3)$	0.88815	0.22
$^1P(4)$	0.88840	
$^3P(1)$	0.86428	34.0
$^3P(2)$	0.88526	6.38
$^1D(1)$	0.8682	32.4
$^1D(2)$	0.88647	4.89
$^1D(3)$	0.88869	0.42
$^3D(1)$	0.88212	7.55
$^3D(2)$	0.88463	0.17
$^3D(3)$	0.88858	0.06
$^1F(1)$	0.88701	0.10
$^3F(1)$	0.8772	2.3
$^3F(2)$	0.88806	1.0

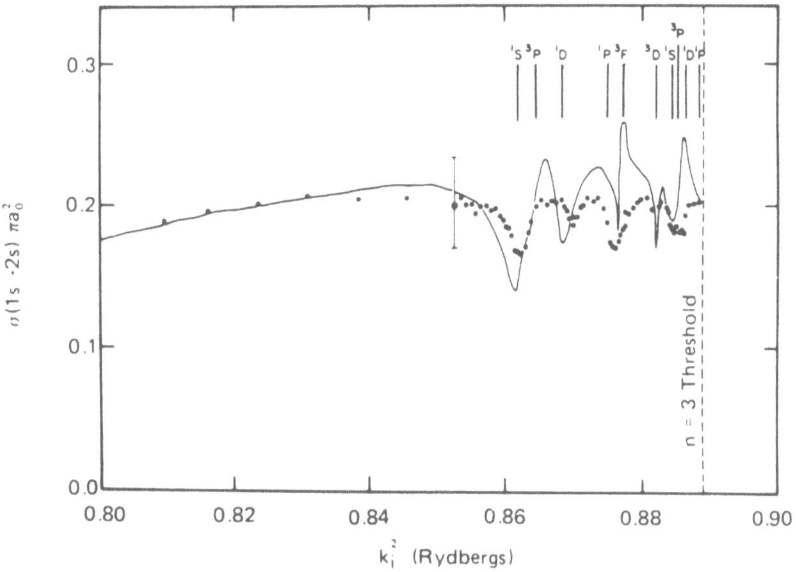

Fig. 6.10 e - H scattering, 1s - 2s excitation cross section (Callaway, 1978): experimental points (Williams, 1976).

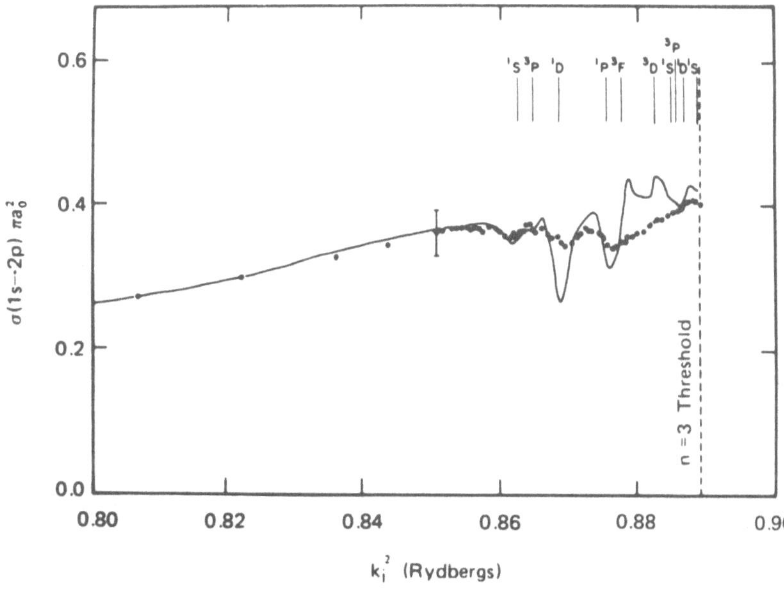

Fig. 6.11 e - H scattering, 1s - 2p excitation cross section (Callaway, 1978): experimental points (Williams, 1976).

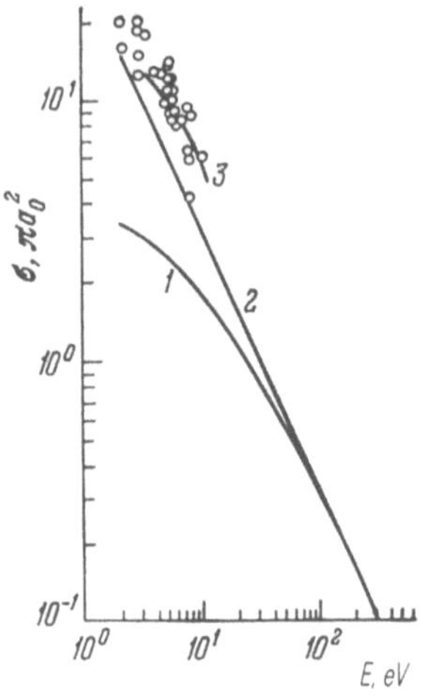

Fig. 6.12 Elastic scattering e - H cross section (Franco, 1968):
1. Born approximation;
2. Glauber approximation;
3. Experimental data.

C. Intermediate and High-Energy Region

Elastic scattering. The elastic scattering cross section was calculated using the Glauber approximation by Franco (1968). In Figure 6.12 the results are presented together with the Born approximation results and experimental data.

Excitation of the 2s and 2p states. The Glauber approximation was applied by Tai et al. (1970). The results are shown in Figures 6.13 and 6.14. The differential cross sections for the n = 2 state excitation (sum of 2s and 2p) at 100 eV and 200 eV are shown in Figures 6.15 and 6.16, together with the experimental results.

Closed Glauber amplitudes for general transitions 1s - nlm in hydrogen were constructed by Thomas and Franco (1976).

Excitation of high-n states - extrapolation procedure. An extrapolation procedure was suggested by McCarroll (1957). At large n we are interested in the excitation cross section for a group of states between n and n + dn rather than some particular n. This cross section can be written as

$$d\sigma = \sigma_n dn = \sigma_n n^3 dE. \tag{1.4}$$

On the other hand the total ionization cross section can be represented in the form

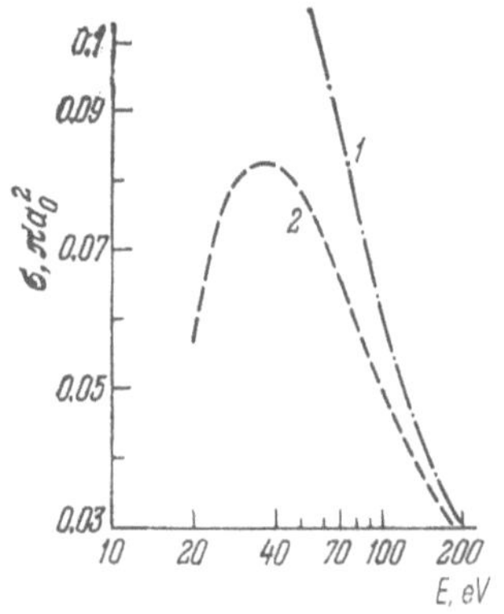

Fig. 6.13 2s-state excitation cross section for H:
1. Born approximation;
2. Glauber approximation (Tai et al., 1970).

$$d\sigma = \sigma_\varepsilon d\varepsilon. \tag{1.5}$$

We can expect that $n^3\sigma_n$ goes over to σ_ε in the limit $n \to \infty$. In Figure 6.17 the results obtained in the Born approximation are shown for two incident electron energies: 123 eV and 220 eV. Negative E corresponds to discrete levels and the curves are drawn for $n^3\sigma_n$. Positive E corresponds to ionization and curves for σ_ε are drawn.

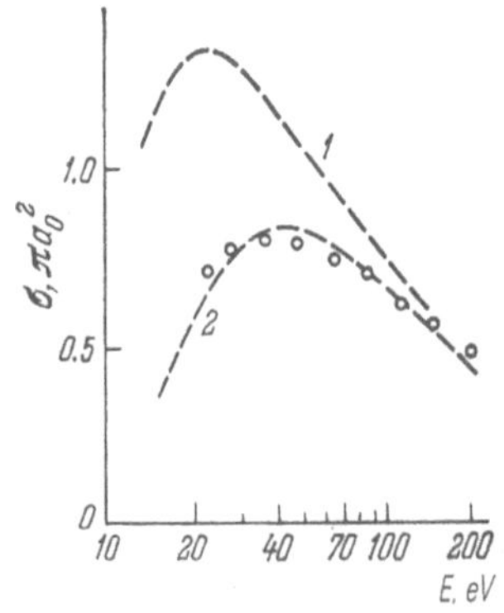

Fig. 6.14 2p-state excitation cross section for H:
1. Born approximation;
2. Glauber approximation;
O Experimental data (Tai et al., 1970).

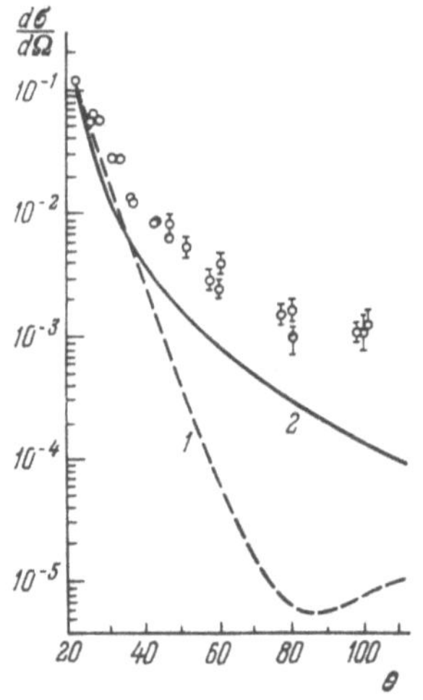

Fig. 6.15 Differential cross section for 2s + 2p excitation, incident electron energy 100 eV:

1. Born approximation;
2. Glauber approximation;

o – experimental data (Tai et al., 1970).

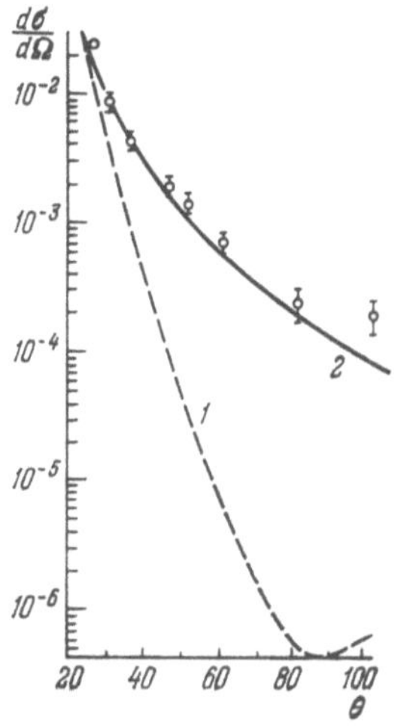

Fig. 6.16 Differential cross section for 2s + 2p excitation, incident electron energy 200 eV:

1. Born approximation.
2. Glauber approximation;

o – experimental data (Tai et al., 1970).

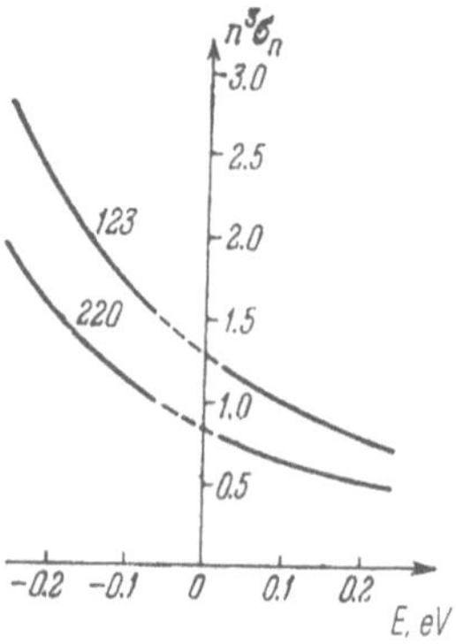

Fig. 6.17 Interpolation of $n^3\sigma_n$ according to McCarroll, 1957. Numbers on the curves indicate the incident electron energy.

An interesting application of this extrapolation procedure is the determination of the energy dependence of σ_n near the threshold. It follows from (1.4) - (1.5) that

$$\sigma_n = \frac{1}{n^3}\frac{d\sigma_\varepsilon}{d\varepsilon} . \qquad (1.6)$$

To determine $\frac{d\sigma_\varepsilon}{d\varepsilon}$ one should use the expression $\sigma_\varepsilon = \text{const}(E - E_0)^\alpha$ (Wannier, 1953; see Section 7.1) where E_0 is the threshold energy, and $\alpha \approx 1.127$. Then

$$\sigma_n = \frac{\text{const}}{n^3}(E - E_0)^{0.127} . \qquad (1.7)$$

The same result was obtained in another way by Vinkaln and Gailitis (1967).

A special feature of high-n excitation near threshold was suggested by Fano (1974): it is accompanied by high ℓ values (though not so high as the maximum possible value n - 1).

The reason is that in the process of excitation the incident electron and the target atom form a transient complex. The incident and excited electrons become closely correlated within the complex due to their mutual Coulomb repulsion which leads to a specific angular momentum distribution. This correlation is very important when both electrons in the final state are slow, which is just the case in high-n excitation near threshold as well as the case of ionization near their threshold (see Section 7.1).

Transitions between high-n states. Various aspects of the transitions in question were reviewed by Percival and Richards (1975). The cross sections summed over ℓ and m in the final state and then averaged over ℓ and m in the initial state have been calculated in the Born approximation by Beigman and Urnov (1974). In the case $n \gg 1$, $\Delta n = |n - n'| \ll n$ and high incident electron energy, there is an approximate analytical expression

$$n^2\sigma_{nn'} = \frac{\pi}{k^2}\left(\frac{nn'}{\Delta n}\right)^3\left(\frac{A}{\Delta n}\ln Ck + B\right) \tag{1.8}$$

where

$$A \simeq 1,\ B \simeq 0.9,\quad C \simeq 0.8(\Delta n)^{-0.07}. \tag{1.9}$$

A similar expression can be obtained using a semiclassical approximation (Beigman, Vainstein and Sobelman, 1969). If the logarithmic term can be neglected then a scaling relation follows:

$$\frac{(\Delta n)^3}{n^4}\,\sigma_{nn'} = \text{const}/k^2. \tag{1.10}$$

6.2 HELIUM

A. Energy Region Below the First Excitation Threshold

A straightforward close coupling calculation including only real bound states and no pseudo-states does not give good results for the elastic scattering cross section. The reason is that for helium 52.4% of the polarizability comes from continuum states (while for hydrogen the contribution from the continuum is less than 20%). So if one takes into account only several discrete states, it leads to a serious underestimation of the polarizability. And in the case of helium the scattering cross section is very sensitive to the value of the polarizability.

To get good results more sophisticated calculations are needed. Two recent calculations can be mentioned in which the target atom electron correlation and the dipole and quadrupole polarizabilities are taken into account: O'Malley et al. (1979) using the R-matrix method and Nesbet (1979) using the matrix variational method. We will present here some details and results of Nesbet's work, and compare with the result of O'Malley et al. and experimental data. Like the case of e - H scattering, phases for $\ell \geq 2$ were computed using expression (1.1), while phases for $\ell = 0$ and $\ell = 1$ were computed by means of the variational method. To calculate the differential cross section expression (1.2) is used. It leads to the following expression for the total cross section (Thompson, 1966)

$$\sigma = \frac{4\pi}{k^2}\sum_{\ell=0}^{2}(2\ell+1)\sin^2\delta_\ell + \frac{\pi^3\alpha^2k^2}{2450}. \tag{2.1}$$

The computed scattering length is equal to 1.18 a_0. In Figure 6.18 a summary of some computed and observed elastic scattering cross sections are shown below 19 eV. The calculations do not include

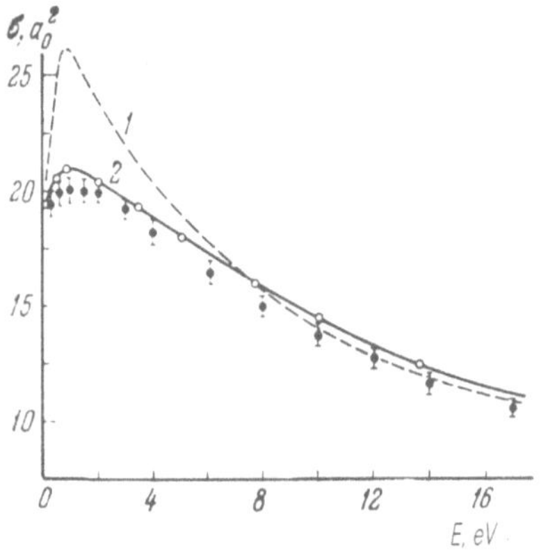

Fig. 6.18 e - H scattering, full line-matrix variational method. Dashed line: close coupling; circles: polarized orbital method (Sinfailam and Nesbet, 1972); ♦ - Experimental data.

the resonance at 19.37 eV. Note that in accordance with equation (5.7) of Chapter 1 $(\frac{d\sigma}{dk})_{k=0}$ is positive. Moreover the numerical value of $(\frac{d\sigma}{dk})_{k=0}$ deduced from the results of Nesbet agrees with expression $\frac{8\pi^2}{3}$ αa, given by (5.7) of Chapter 1. In Figure 6.19 the differential cross sections for several energies are compared with the measured angular distributions of Andrick and Bitsch (1975) scaled by the ratio of the calculated value of σ to their published value.

Turning to the resonances below the first excitation threshold, one should note that for helium there is just one resonance instead of infinite series as in the case of hydrogen. This resonance is due to a ^2S state of He^-, and it is located at 19.37 eV. The resonance profile has been measured by Kennerly et al. (1981) with a very high energy resolution, $5\cdot10^{-3}$ eV. The measured width of this resonance is $(11.0 \pm 0.5) \times 10^{-3}$ eV.

The measured values are in good agreement with recent calculations (Hazi, 1978; Junker, 1978; Foster et al., 1979). Just at the 2^3S threshold there is a rounded step structure in the scattering cross section discussed in Section 3.3. This rounded step has been observed by Cvejanovic et al. (1974) (Figure 6.20).

B. Region between the n = 2 and n = 3 Thresholds

There are 4 states in this region: 2^3S (19.82 eV), 2^1S (20.61 eV), 2^3P (20.96 eV) and 2^1P (21.22 eV).

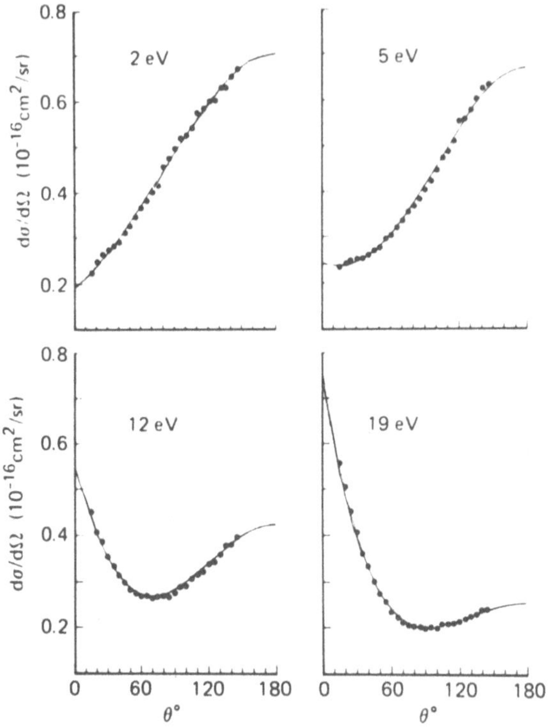

Fig. 6.19 Differential cross section for e - He elastic scattering. Theory: Nesbet (1979); experiment: Andrick and Bitsch (1975).

Theory and experiment in this region were recently reviewed by Nesbet (1980), and we will give here only a concise summary of the results. The rather complicated energy dependence of the cross sections in this region can be interpreted in terms of the sum of S-matrix eigenphases for the three states ^2S, $^2P^0$ and ^2D of the He + e system. The eigenphase sums are plotted in Figure 6.21

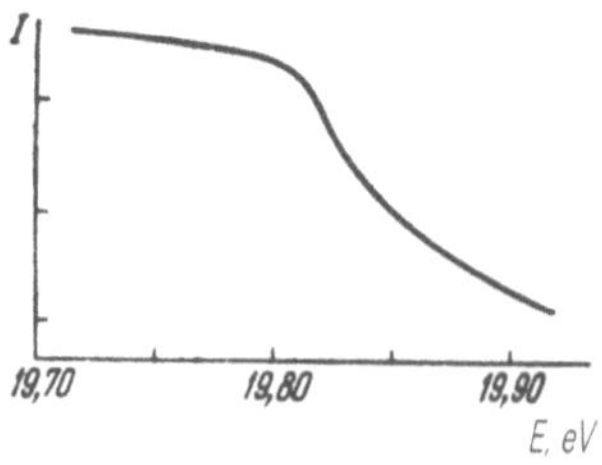

Fig. 6.20 e - He scattering, rounded step at the 2^3S threshold (Cvejanovic et al., 1974).

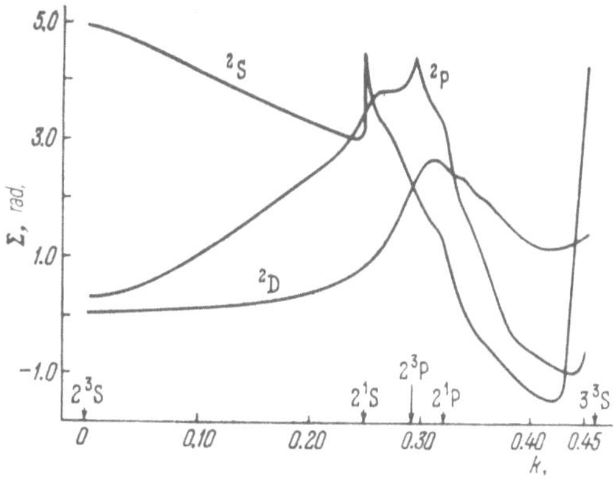

Fig. 6.21 e - He scattering, S matrix eigenphase sums (Oberoi and Nesbet, 1973).

against electron momentum in the 2^3S channel. The ^2S eigenphase sum descends from the 2^3S threshold, following its rise through π at the narrow ^2S resonance below the 2^3S threshold. At the 2^1S threshold the ^2S eigenphase sum rises rapidly by $\simeq \pi/2$ and then decreases. The ^2P eigenphase sum rises through π in a broad resonance below the 2^1S threshold and shows a cusp structure at the 2^3P threshold and a rounded step structure at the 2^1P threshold. The ^2D eigenphase sum shows a resonance near the 2^3P threshold.

Elastic scattering on the 2^1S and 2^3S states was computed by Burke et al. (1969a) and Oberoi and Nesbet (1973). The results together with experimental data (Neynaber, 1964) are shown in Figures 6.22 and 6.23. The threshold behavior of the cross sections is strongly affected by the very large polarizability of the He atom in excited states: 316 for the 2^3S and 802 for the 2^1P states.

The excitation cross sections for the 2^3S and 2^1S states are shown in Figures 6.24 and 6.25 with computed values (Oberoi and Nesbet, 1973; Burke et al., 1969a) and experiment (Brongersma et al., 1972). There is a large uncertainty in the absolute normalization of the experimental cross section indicated by the error bar. If the experimental data were normalized to the variational calculation at some point, the two curves would be in close agreement.

A special investigation of the threshold behavior of the excitation cross sections (Nesbet, 1975) leads to the following results:

a) the 1^1S - 2^1S cross section has a very sharp peak with a width of 5.10^{-3} eV (Figure 6.26);
b) the 1^1S - 2^3S cross section varies rather smoothly near threshold. For comparison both cross sections are shown in Figure 6.27 on a common energy scale.

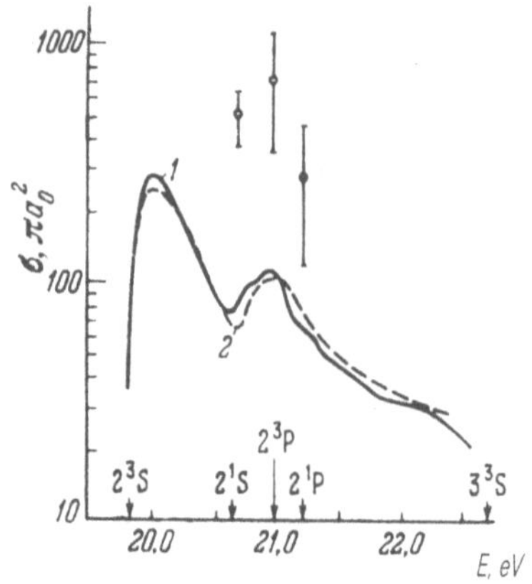

Fig. 6.22 Elastic scattering on He(2^3S). Theory:
1. Oberoi and Nesbet (1973);
2. Burke et al. (1969a).
Experiment: Neynaber (1964).

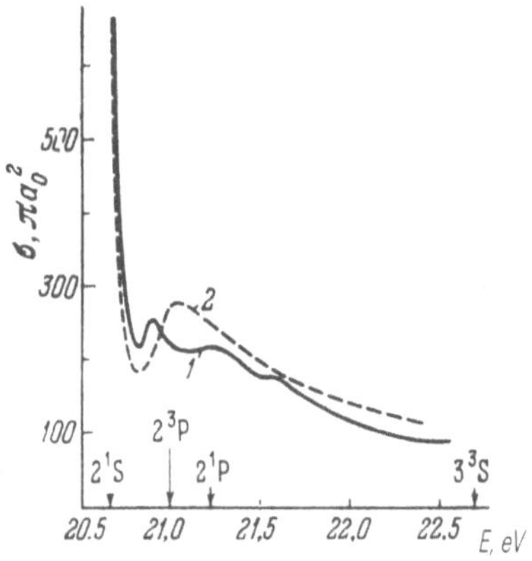

Fig. 6.23 Elastic scattering on He(2^1S). Theory:
1. Oberoi and Nesbet (1973);
2. Burke et al. (1969a).
Experiment: Neynaber (1964).

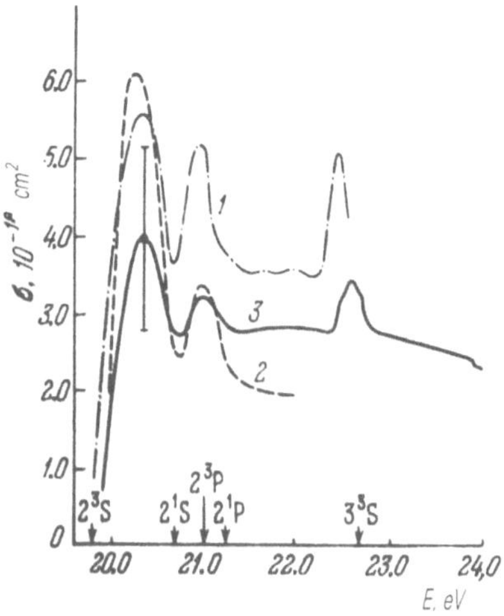

Fig. 6.24 e - He scattering, 2^3S state excitation cross section.
Theory: 1. Oberoi and Nesbet (1973);
2. Burke et al. (1969a).
Experiment: Brongersma et al. (1972).

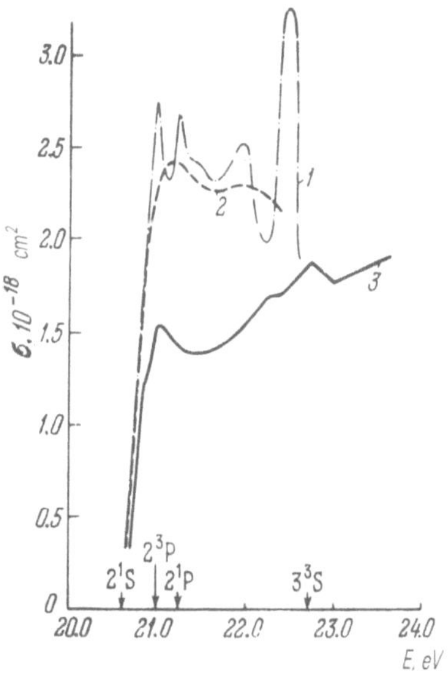

Fig. 6.25 e - He scattering, 2^1S state excitation cross section.
Theory: 1. Oberoi and Nesbet (1973);
2. Burke et al. (1969a).
Experiment: Brongersma et al. (1972).

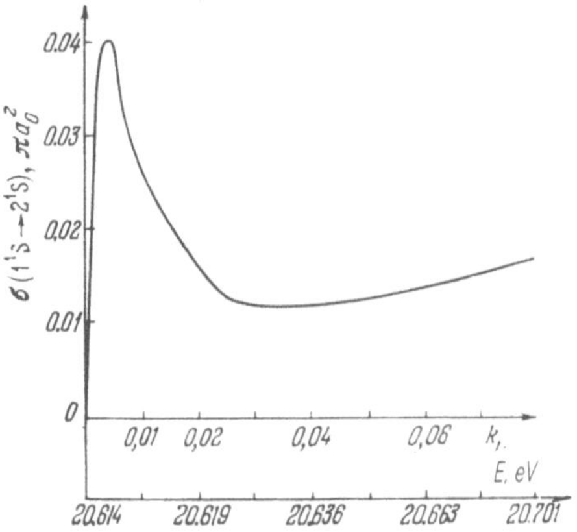

Fig. 6.26 e - He scattering, 2^1S state excitation cross section near the threshold (Nesbet, 1975).

c) the ratio

$$R = \frac{\int_0^{\Delta E} \sigma(2^1S)dE}{\int_0^{\Delta E} \sigma(2^3S)dE}$$

is very sensitive to the value of ΔE as shown in Figure 6.27.

All this agrees well with high-resolution experiments of helium threshold excitation (Cvejanovic and Read, 1974). The n = 3 states were not represented in the calculations of Oberoi and Nesbet (1973). More refined calculations which include the n = 3 states were carried out by Nesbet (1978).

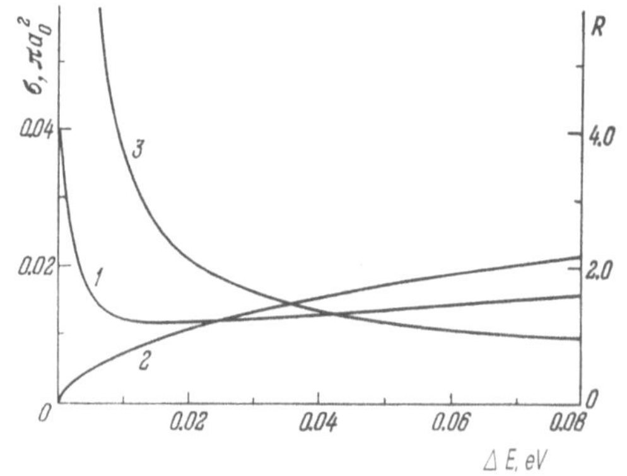

Fig. 6.27 e - He scattering:
1. 2^1S state excitation cross section;
2. 2^3S state excitation cross section;
3. R (ΔE) (Nesbet, 1975).

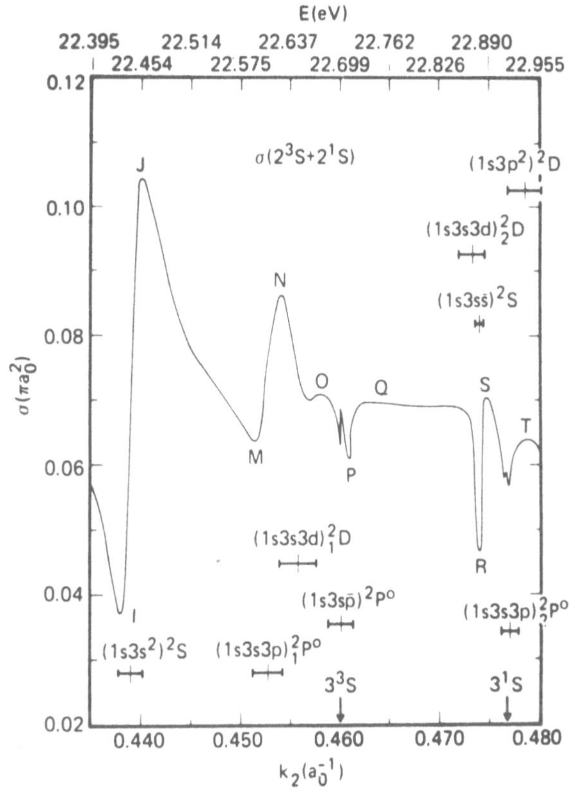

Fig. 6.28 e - He scattering, computed metastable excitation cross section (Nesbet, 1978): Computed resonance positions and widths are indicated by vertical and horizontal bars and identified by symbols; structural features of the curve are indicated for comparison with experiment (Figure 6.29).

The computed total cross section for excitation of the 2^3S and 2^1S states is shown in Figure 6.28, k_2 being the momentum relative to the 2^3S threshold. The measured cross section (Brunt et al., 1977) is shown in Figure 6.29. Comparison of Figures 6.28 and 6.29 indicates close correspondence of all the calculated and measured structural features. The computed resonance positions and widths are shown in Figure 6.28 by small vertical and horizontal bars together with corresponding configurations and spectroscopic symbols.

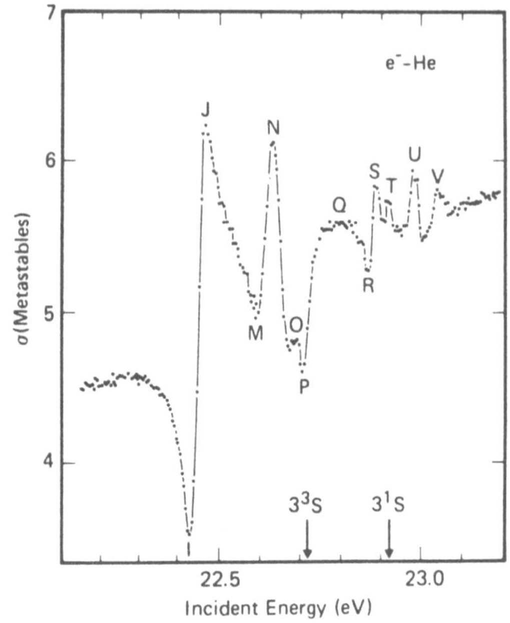

Fig. 6.29 e - He scattering, observed metastable excitation cross section (Brunt et al., 1977).

C. Intermediate and High-Energy Region

The elastic-scattering differential cross section was computed in the Glauber approximation by Franco (1970). In Figure 6.30 the results for 100 and 150 eV are shown together with the Born approximation and experimental data. The differential cross section is plotted against the square of the momentum transfer q^2:

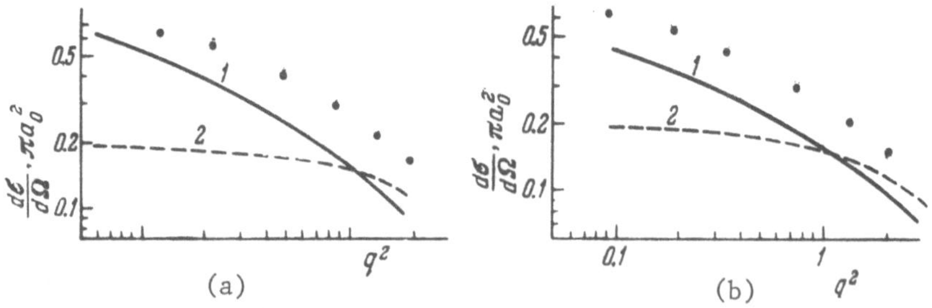

Fig. 6.30 e - He scattering, differential elastic cross section: 1. Glauber approximation; 2. Born approximation; ● - experimental data; Incident electron energy: (a) 100 eV; (b) 150 eV (Franco, 1970).

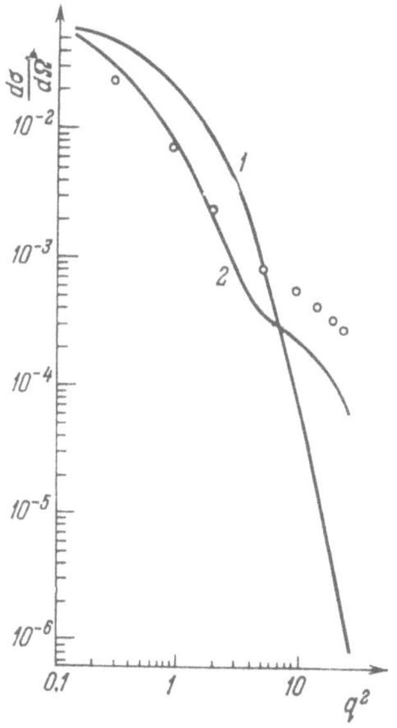

Fig. 6.31 e - He scattering, differential cross section for 2^1P state excitation at an incident electron energy 100 eV: 1. Born approximation; 2. Glauber approximation; o – experimental data (Franco, 1970).

$$q = 2k \sin\theta/2.$$

The 2^1P excitation differential cross section was also computed in the Glauber approximation for a number of energies by Franco (1973). As an example the cross section for 100 eV is shown in Figure 6.31.

6.3 ALKALI METAL ATOMS (Li, Na, K, Cs)

A. Energy Region below the First Excitation Threshold

Let us consider first some special features of the alkali metal atoms. In alkali metals the first excited state gives more than 90% of the atomic polarizability. Thus one can expect that the close coupling calculation which takes into account only the first excited state (the two-state approximation) will be quite accurate.

The alkali metal atoms can be considered as a good approximation to one-electron systems (valence electron in the field of a core). Hence, the scattering problem is similar to that by a hydrogen-like system.

All alkali metal atoms have stable negative ions. The binding energy of the negative ions is rather low: 0.609 eV for Li^-, 0.548 eV for Na^-, 0.501 eV for K^-, 0.471 eV for Cs^-, 0.4859 eV for Rb^-. Hence, one can expect a large value of the scattering length. But the very high polarizability (164 for Li, 159 for Na, 293 for K) leads to a significant change of the scattering length.

In Table 3 the scattering lengths computed by Karule (1965) in the two-state close coupling approximation, a, and in one-state approximation, a_0 (no coupling with excited states), are listed. Finally, there are quasistationary ^3P states of negative ions with very low energies:

$$Li^-: E = 6.0 \times 10^{-2} \text{ eV}, \Gamma = 5.7 \times 10^{-2} \text{ eV}$$

$$Na^-: E = 8.3 \times 10^{-2} \text{ eV}, \Gamma = 8.5 \times 10^{-2} \text{ eV}$$

$$K^- : E = 2.4 \times 10^{-3} \text{ eV}, \Gamma = 5.8 \times 10^{-4} \text{ eV}$$

(Sinfailam and Nesbet, 1973). One can expect that the $\ell = 1$ partial wave should dominate at low energy so the angular distribution should be very anisotropic. Now we turn to the calculations. There are two-state close coupling calculations (Karule, 1965, 1972; Burke and Taylor, 1969b; Norcross, 1971); four-state close coupling calculations for Na, extended into the inelastic region (5 eV) (Moores and Norcross, 1972); and variational calculations (Sinfailam and Nesbet, 1973).

In Figures 6.32, 6.33, 6.34, 6.35, 6.36 and 6.37 the computed phases are shown as $\ell = 0, 1, 2$ for Li and $\ell = 0, 1, 2$ for Na. In Figures 6.38 and 6.39 the spin-averaged elastic cross section for Li and K are shown. The experimental data are taken from Perel et al. (1962). Note that the ^1P and ^1D phases rise sharply below the n_0P excitation threshold. This behavior can be interpreted in terms of a ^1D resonance and a cusp superimposed on a ^1P resonance (Bardsley and Nesbet, 1973).

Table 3. Scattering Lengths for Alkali Metal Atoms

	a_0		a	
	Singlet	Triplet	Singlet	Triplet
Li	18.8	5.92	3.65	- 5.66
Na	17.5	5.69	4.23	- 5.91
K	18.1	6.71	0.45	- 15
Cs	16.6	7.59	- 4.04	- 25.3

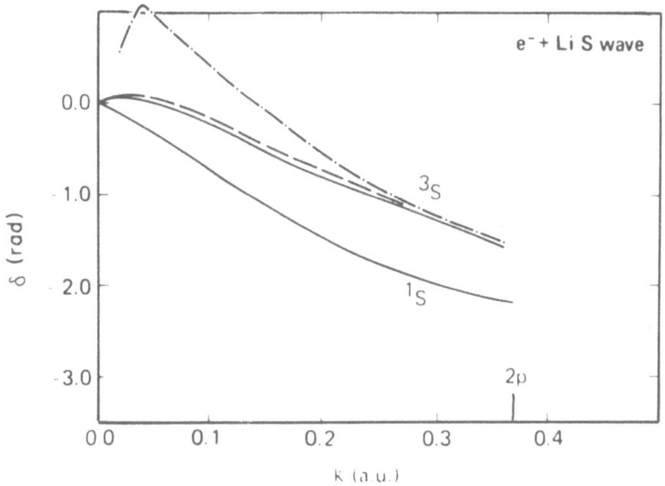

Fig. 6.32 e - Li scattering, ℓ = 0 singlet and triplet phases (reproduced from Nesbet, 1980):
—— Sinfailam and Nesbet (1973);
-·- Burke and Taylor (1969b);
--- Norcross (1971).

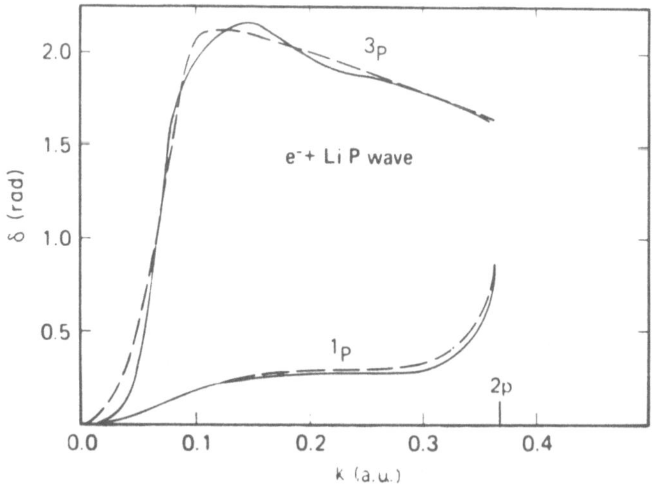

Fig. 6.33 e - Li scattering, ℓ = 1 singlet and triplet phases (reproduced from Nesbet, 1980):
—— Sinfailam and Nesbet (1973);
--- Burke and Taylor (1969b).

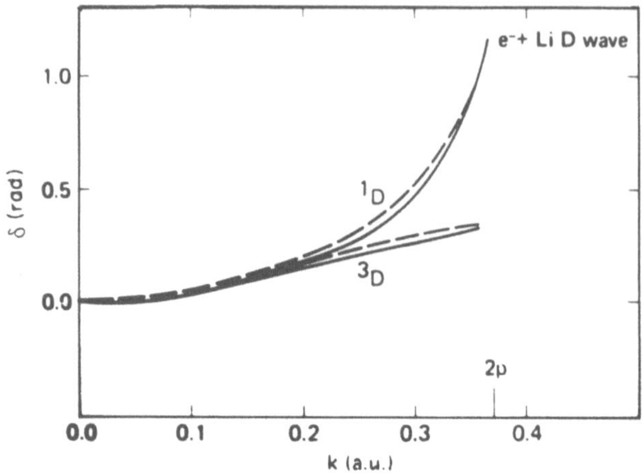

Fig. 6.34 e - Li scattering, ℓ = 2 singlet and triplet phases (reproduced from Nesbet, 1980):
—— Sinfailam and Nesbet (1973);
--- Burke and Taylor (1969b).

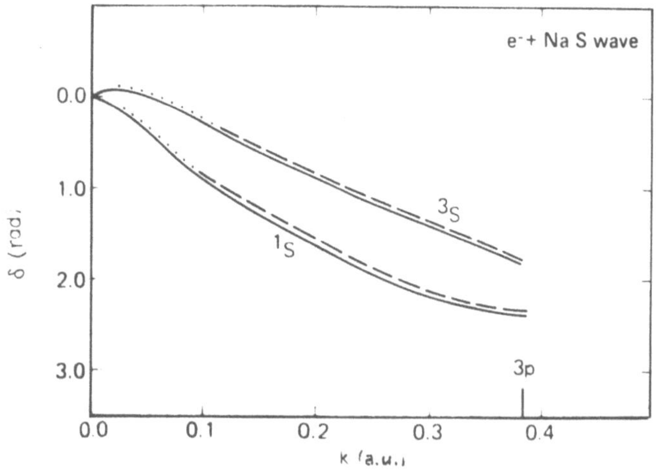

Fig. 6.35 e - Na scattering, ℓ = 0 singlet and triplet phases (reproduced from Nesbet, 1980):
—— Sinfailam and Nesbet (1973);
--- Moores and Norcross (1972);
... Norcross (1971).

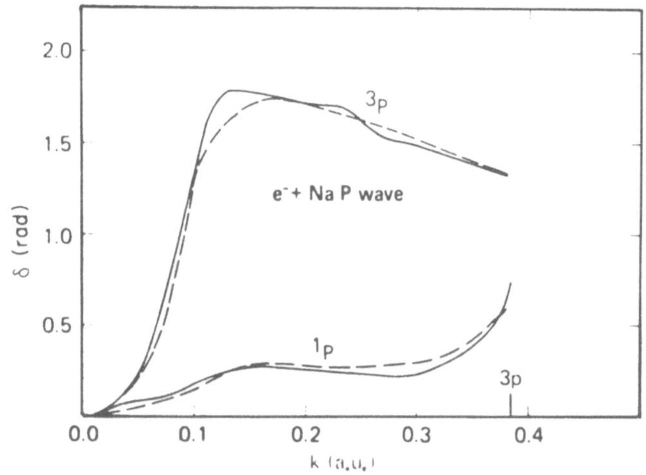

Fig. 6.36 e - Na scattering, ℓ = 1 singlet and triplet phases (reproduced from Nesbet, 1980):
—— Sinfailam and Nesbet (1973);
--- Moores and Norcross (1972).

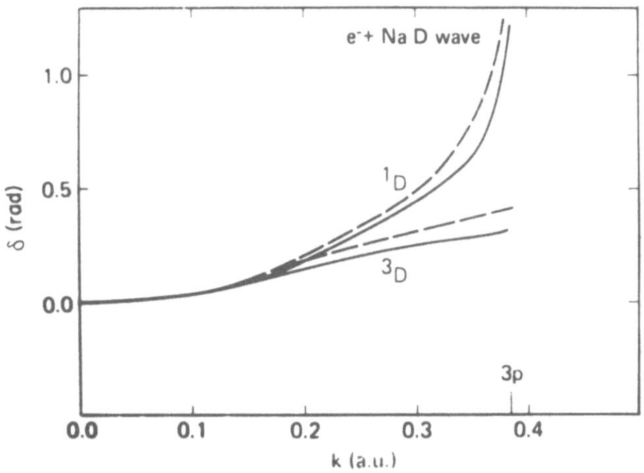

Fig. 6.37 e - Na scattering, ℓ = 2 singlet and triplet phases (reproduced from Nesbet, 1980):
—— Sinfailam and Nesbet (1973);
--- Moores and Norcross (1972).

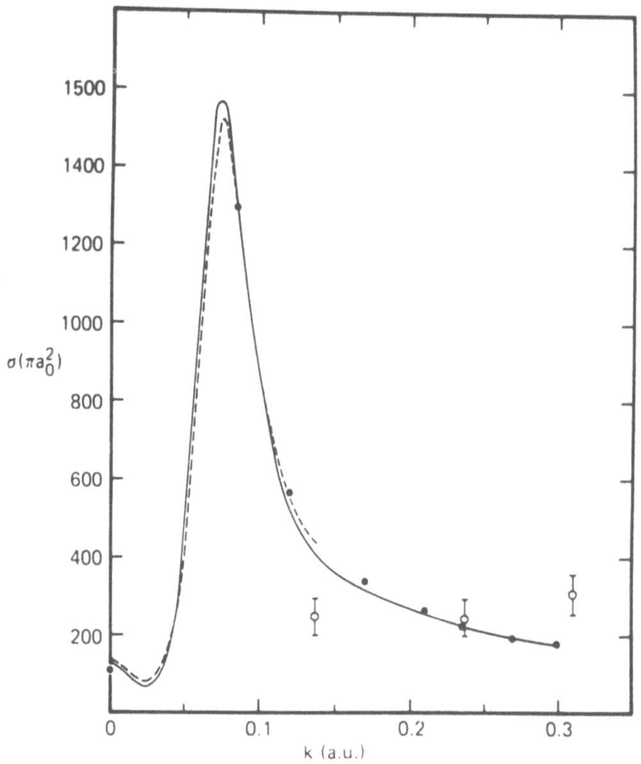

Fig. 6.38 e - Li scattering, spin-averaged elastic cross section.
Theory: —— Sinfailam and Nesbet (1973);
--- Norcross (1971);
... Karule (1965).
Experiment: ϕ Perel et al. (1962).

In Figure 6.40 the spin-averaged elastic scattering differential cross section for Na is shown at several energies. It is highly anisotropic as was expected. Note the steep slope at $\theta = 0$ due to the large polarizability and large scattering length.

Elastic scattering differential cross sections for Li, Na, K and Cs were computed by Karule (1972) in the singlet and triplet states separately. The results were used to calculate the spin-averaged total elastic scattering cross section and to calculate the electron spin polarization after scattering on a polarized atomic beam. The high degree of polarization of the electrons scattered by a polarized beam of alkali metal atoms at very low energy was first suggested by Drukarev and Ob'edkov (1971). The polarization was estimated using the scattering length values in this paper. Systematic calculations performed by Karule (1972) confirmed this earlier suggestion.

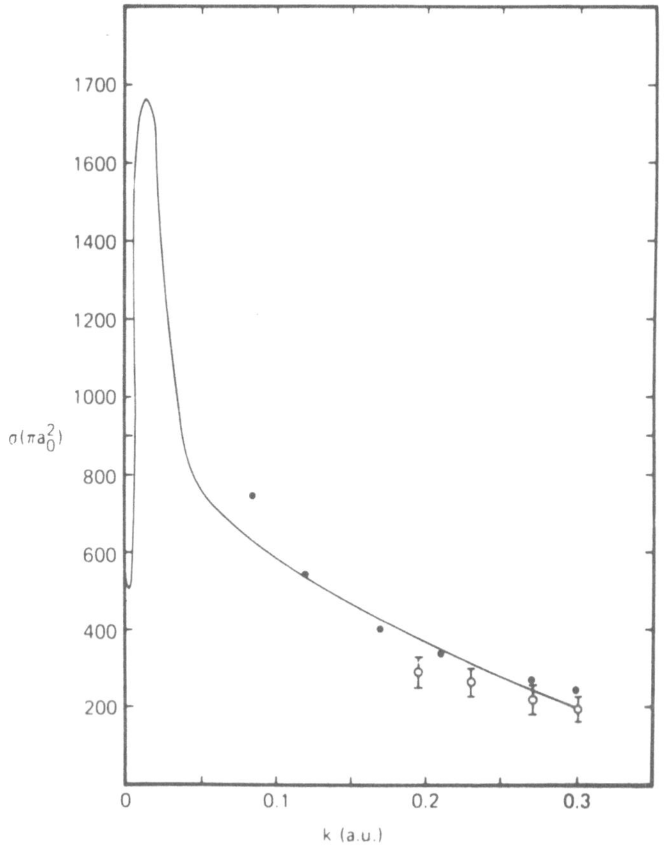

Fig. 6.39 e - K scattering, spin-averaged elastic cross section.
Theory: —— Sinfailam and Nesbet (1973);
... Karule (1965).
Experiment: ϕ Collins et al. (1971).

B. Region above the First Excitation Threshold

The elastic scattering differential cross sections for Li, Na, K and Cs were computed for several energies by Karule and Peterkop (1972). As an example the differential cross section for K at 3 eV is shown in Figure 6.41 together with the experimental data in arbitrary units. Only the shape of the experimental and theoretical curves can be compared.

The values of the total spin-averaged elastic scattering cross section for several energies calculated in this paper are listed in Table 4.

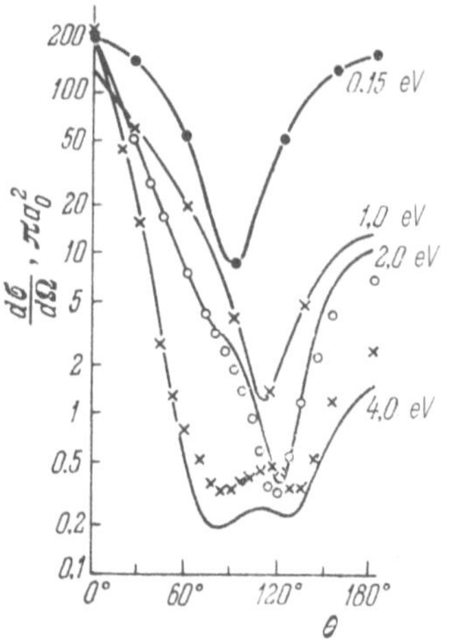

Fig. 6.40 e - Na scattering differential cross section, Moores and Norcross (1972);
●,○,x – two-state close coupling.

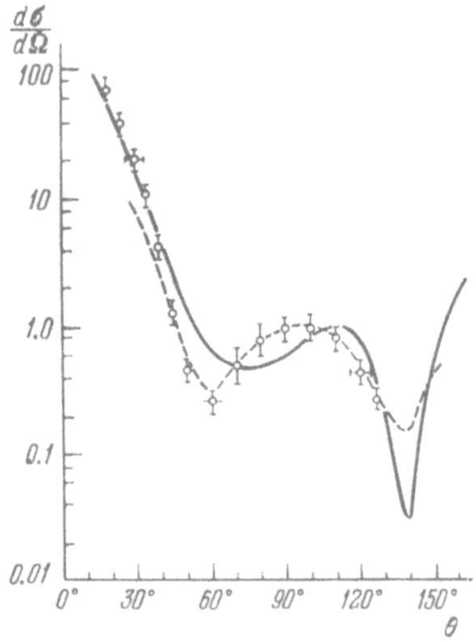

Fig. 6.41 e - K scattering differential cross section at 3 eV, Karule and Peterkop (1972);
various experimental results in arbitrary units are shown.

Table 4. Elastic Scattering Cross Section below the Excitation Threshold

E,eV	$\sigma_{Li}, \pi a_o^2$	$\sigma_{Na}, \pi a_o^2$	$\sigma_K, \pi a_o^2$	$\sigma_{Cs}, \pi a_o^2$
0	109	122	672	1937
0.1	1302	694	748	448
0.2	582	709	538	499
0.3	--	503	--	--
0.4	341	396	405	501
0.6	265	289	336	492
0.8	223	233	297	483
1.0	196	198	271	467
1.2	181	175	252	442
1.4	165	157	240	434
1.6	158	145	230	--
1.8	159	136	--	--
2	--	133	--	--

Excitation cross sections for the first P-excited states for Li, Na, K and Cs were computed by Karule and Peterkop (1971), and for Na also by Moores and Norcross (1972). The values of the excitation cross sections at several energies are listed in Table 5, taken from Karule and Peterkop (1971). Finally, we consider the total elastic + inelastic scattering cross sections for Na, which were measured by Kasdan et al. (1973). Absolute values of the total cross section were obtained in this experiment with a rather low error. In Figure 6.42 the experimental data are compared with two- and four-state close coupling calculations. Note a very distinct cusp structure at the threshold.

Table 5. Excitation Cross Section for P states

E,eV	$\sigma_{Li}, \pi a_o^2$	$\sigma_{Na}, \pi a_o^2$	$\sigma_K, \pi a_o^2$	$\sigma_{Cs}, \pi a_o^2$
1.6	--	--	--	40
1.8	--	--	15	29
2	10	--	31	27
2.5	--	16	--	--
3	33	27	50	53
4	44	33	61	68
5	44	38	66	73

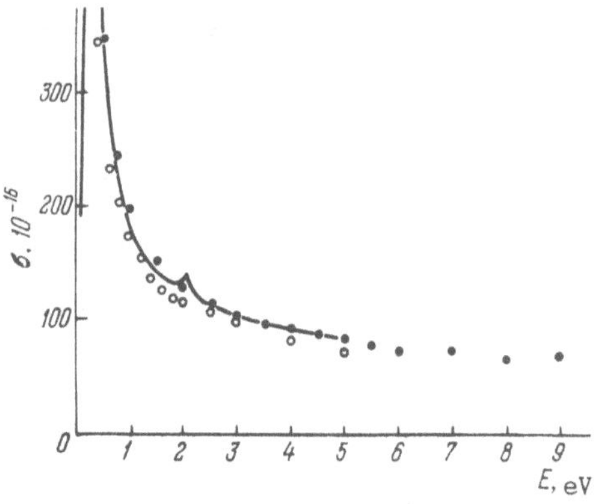

Fig. 6.42 e - Na scattering, total elastic + inelastic cross section.
Theory: Moores and Norcross (1972);
o – Karule and Peterkop (1965);
• – experiment (Kasdan et al., 1973).

6.4 OTHER ATOMS: REFERENCES AND COMMENTS

Alkaline Earth Atoms (Be, Mg, Ca, Sr, Ba)

These atoms have two valence electrons. If the action of the other electrons and the nucleus on the incident electron could be replaced by an effective potential, we would have a helium-like collision problem. But unlike helium, the lowest excited states are ^3P and ^1P for Be and Mg, and ^3D and ^1D for Ca, Sr, Ba. The lowest excited states give the most important contribution to the polarizability (85% for Be). Thus, one can expect comparatively good results from two- or three-state close coupling calculations. In fact the simplified version of the close coupling calculation based on the helium-like model was performed by Van Blerkom (1970) and Fabrikant (1975). It was shown by Fabrikant that the results are rather sensitive to a particular selection of coupled states.

For instance, the elastic scattering cross section for Ba calculated in ^1S - ^1P close coupling approximation differs significantly from that in ^1S - ^1P - ^3P close coupling approximation (Figure 6.43).

The important factor in such calculations is the interelectronic correlations of the valence electrons in the ground and excited states.

Fabrikant used Hartree-Fock wave functions for Be and semi-empirical wave functions for other atoms. In both, the interelectronic correlation is neglected completely.

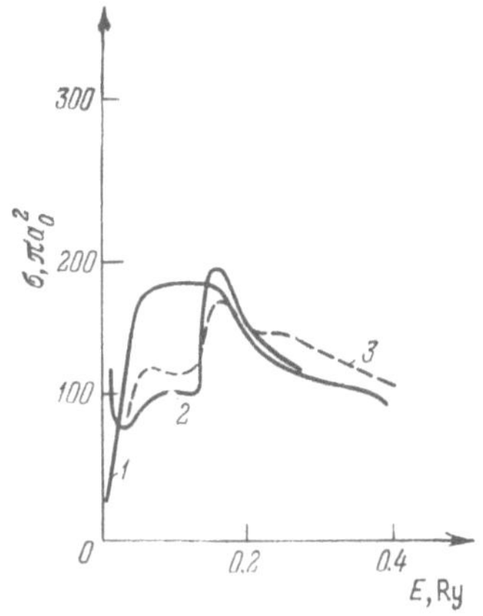

Fig. 6.43 e - Ba elastic scattering cross section (Fabrikant, 1975):
1. ^1S - ^1P state close coupling;
2. ^1S - ^1P - ^3P state close coupling;
3. effect of D states.

However, the correlation correction for example, calculated by means of configuration interaction, changes the polarizability of Be by a factor of 1.5.

Resonances in elastic scattering have been found by Fabrikant, their positions being 0.136 eV in Be, 2.584 eV in Mg, 1.768 eV in Ca, 0.952 eV in Sr and 2.04 eV in Ba. The angular momentum quantum numbers are 1 for Be, Ca, Ba and 2 for Mg and Sr.

The resonances in elastic scattering on Be, Mg, Zn and Cd were recently considered by Sinfailam (1981).

Carbon, Nitrogen and Oxygen Atoms

These atoms have the following ground state configurations:

$$(1s)^2(2s)^2(2p)^2 \text{ for C,}$$
$$(1s)^2(2s)^2(2p)^3 \text{ for N,}$$
$$(1s)^2(2s)^2(2p)^4 \text{ for O.}$$

The possible terms corresponding to these configurations are: ^1S, ^1D, ^3P for C and O. The ^3P term is the lowest, ^1S and ^1D being the excited states. For N the possible terms are ^2P, ^2D, ^4S. The ^4S term is the lowest, ^2P and ^2D being the excited states.

In addition, the excited ^5S state in C, belonging to the configuration $(1s)^2(2s)(2p)^3$, has a small excitation energy and should be considered in close coupling calculations. There are also stable negative ions. C^- has two stable states: one with binding energy 1.46 eV and another with binding energy 0.035 eV. O^- has binding energy 1.27 eV. N^- probably does not exist in a stable form, but there is a low lying resonance state $(2s)^2(2p)^4$ ^3P at about 0.1 eV above the ground state of the system e + $N(^4S)$. The variational calculations of low energy electron scattering on C, N and O were reported in a series of papers by Nesbet et al. and were reviewed recently by Nesbet (1980).

The results depend very strongly upon the details of the wave functions used in the calculations. For instance, in nitrogen if a bound state $N^-(^3P)$ exists, the elastic scattering cross section would be small and varying rather smoothly. But, if instead a resonance exists, then the cross section would have a typical resonance behavior.

To get reliable results a very extended wave function basis should be employed. It should give reasonably accurate ground and low lying excited states, should contain the main contribution to the atomic polarizability, and should ensure the adequate description of the most important resonances. It seems that the present situation in this respect is far from satisfactory.

6.5 COLLISIONS OF ELECTRONS WITH POSITIVE IONS

A. Elastic Scattering below the Excitation Threshold

In the small scattering angle region the effect of the Coulomb tail of the potential dominates leading to the $(k \sin\theta/2)^{-4}$ Rutherford law. In the backscattering region at $\theta = 180°$, the scattering deviates most strongly from the Rutherford law. According to the expression (6.15) of Chapter 1

$$\left(\frac{d\sigma}{d\Omega}\right)_{\theta=\pi} = \frac{Z^2}{4k^4} + \frac{|f|^2}{k^2} + \frac{Z}{k^3}\,\mathrm{Re}\, f.$$

The function f is determined by (6.10), (6.11) and (6.12) of Chapter 1 and can be represented in the form

$$f = \frac{1}{2i}\sum_{\ell}(2\ell+1)e^{i(\phi_\ell+\mu_\ell)}\sin\mu_\ell \tag{5.1}$$

where

$$\phi_\ell = 2\sum_{m=1}^{\ell}\arctan\frac{mk}{Z}\,. \tag{5.2}$$

and μ_ℓ is the additional phase shift due to the action of the ionic core.

At very low energy the values of μ_ℓ can be determined by means of interpolation of the quantum defect [(6.9) of Chapter 1]. However, this estimation becomes less applicable if the energy is increased so that direct computation is needed. It was shown in Chapter 1 that μ_ℓ remains finite even at zero energy. Thus we can expect that several partial waves contribute at low energy. As an example e - Na^+ scattering can be considered (Drukarev and Berezina, 1975). In Figure 6.44 $(\frac{d\sigma}{d\Omega})_{\theta=\pi}$ is plotted against k. Phases up to μ_4 were taken into account. Two sets of μ_ℓ were used. In the first, all phases μ_0 ... μ_4 were calculated by means of quantum defect extrapolations. In the second, phases μ_0, μ_1, μ_2 were calculated using Hartree-Fock wave functions. There is a Ramsauer-like minimum near k = 0.6. In other cases maximum instead of minimum can be obtained, for instance in Sc^{3+} (Figure 6.44b).

It is very instructive to observe the change of μ_ℓ as a function of the ionic charge Z at zero energy (Figure 6 .45). The rapid increase of approximately π can be observed for Z near the beginning of a new electronic shell in the ion. When the energy of the incident electron increases towards the excitation threshold resonances appear. Again, the angular region where resonances are most prominent is near $\theta = 180^\circ$. It follows from consideration of the most simple two-channel system (Section 3.5) that there will be a hydrogen-like infinite series of resonances. At each resonance strong back-scattering can be expected.

B. Excitation of Positive Ions. General Remarks

It is customary in the literature to use the collision strength Ω instead of the cross section whenever an ion excitation is considered. Ω is related to the cross section σ by

$$\Omega = \omega k^2 \sigma \quad (5.3)$$

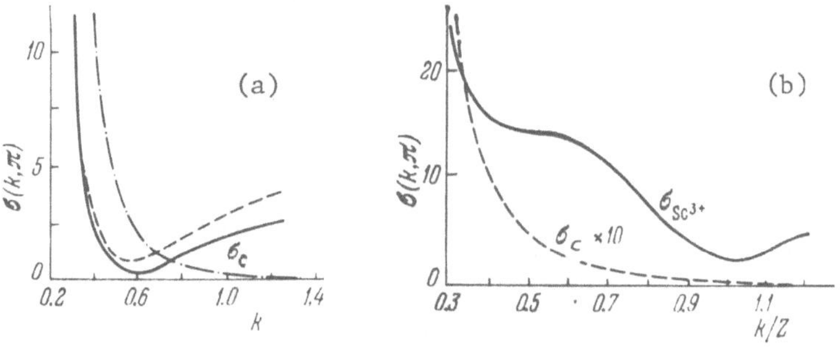

Fig. 6.44 a) e - Na^+ backscattering cross section, σ_c - Rutherford.
b) e - Sc^{3+} backscattering, σ_c - Rutherford.

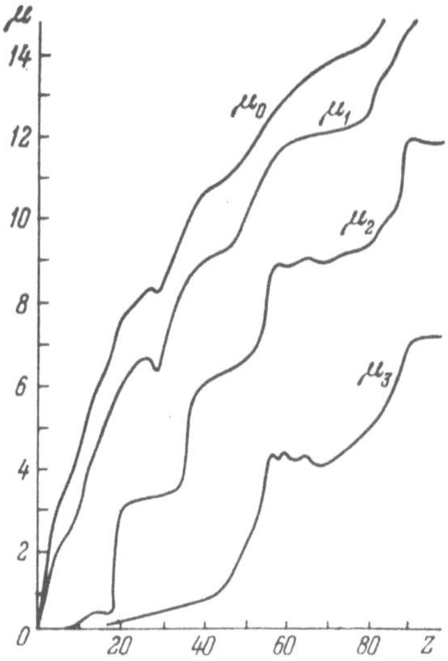

Fig. 6.45 μ_ℓ as a function of Z at zero energy (Manson, 1969).

where $1/2k^2$ is the energy of the incident electron and ω is the statistical weight of the initial state. It is equal to $(2S_i + 1) \times (2L_i + 1)$ for the L - S coupling approximation or $(2J_i + 1)$ if the fine structure of levels is taken into account.

Unlike the case of neutral atoms, the cross section and Ω for ions is finite at the threshold. Resonances play an important role in the region between the first excitation threshold and the ionization threshold. In the case of a positive atomic ion of charge Z the resonances correspond to an excited state of an ion of charge Z - 1. This state has several channels of decay:

1) back to the ion with charge Z in an excited state and a free electron;
2) radiative decay of the resonance state into a stable state of the ion with charge Z - 1, the electron being trapped. This process is called dielectronic recombination. Because the transition probability is proportional to the 4th power of the ionic charge, this channel is especially important for multicharged ions;
3) ejection of an additional electron if it is allowed by energy conservation. This process is of special importance in the case of inner shell excitation.

In an energy region where the resonances lie very close together we are interested in the cross section averaged over the resonances, rather than the cross section itself. If the separation of the resonances is large compared with the widths of the resonances, then:

$$\Omega(i,f) = \Omega^{>}(i,f) + \sum_{i'} \frac{\Omega^{>}(i,i')\Omega^{>}(i',f)}{\sum_{i''} \Omega^{>}(i',i'')} \tag{5.4}$$

where $\Omega^{>}$ are collision strengths calculated above the new threshold and extrapolated to energies below this threshold. In (5.4), i' is summed over the degenerate closed channels of the new threshold and i" is summed over all open channels (Gailitis, 1963). In many applications the quantity of interest is the rate coefficient, $\langle v\sigma \rangle$, where v is the electron velocity and the average is over a Maxwellian distribution. Of course, all the near-threshold resonances in Ω are averaged by integration. But, in spite of averaging, resonances can significantly increase the rate coefficient. A systematic review of excitation of atomic positive ions by electrons including an extended bibliography was given recently by Henry (1981).

C. Collision Strengths near Threshold. Resonances

To give an example of collision strength calculations near threshold we consider the 1s - 2s and 1s - 2p transitions in He^{+} in the vicinity of the threshold. The calculation of the 1s - 2s excitation cross section is of particular interest because there is a still unresolved discrepancy of a factor ~ 2 between the best theoretical results and experiment.

There are several close coupling calculations. We reproduce here the results of recent variational calculations (Morgan, 1979) together with the results of Burke et al. (1964) (three target states), Burke and Taylor (1969a) (three target states and twenty correlation terms), and experimental results for 1s - 2s (Dolder and Peart, 1973). Figure 6.46 shows the results for 1s - 2s, and Figure 6.47, for 1s - 2p. Comparing the results of different calculations, we can see how they are converging as the number of states taken into account increases.

The resonances in the 1s - 2s and 1s - 2p transitions of hydrogen-like ions were recently considered by Hayes and Seaton (1978). The positions of resonances in series converging to the n = 3 threshold are estimated approximately in this paper by assuming that one has an inner electron moving in the field of a charge Z and an outer electron moving in the field of a charge Z - 1. The energy of the (n,n') resonance is then

$$E(n,n') = -\frac{1}{2}\left(\frac{Z}{n}\right)^2 - \frac{1}{2}\left(\frac{Z-1}{n'}\right)^2. \tag{5.5}$$

For excitation to the n = 2 level the excitation energy is

$$\Delta E = \frac{1}{2}\cdot\frac{3}{4}\,Z^2.$$

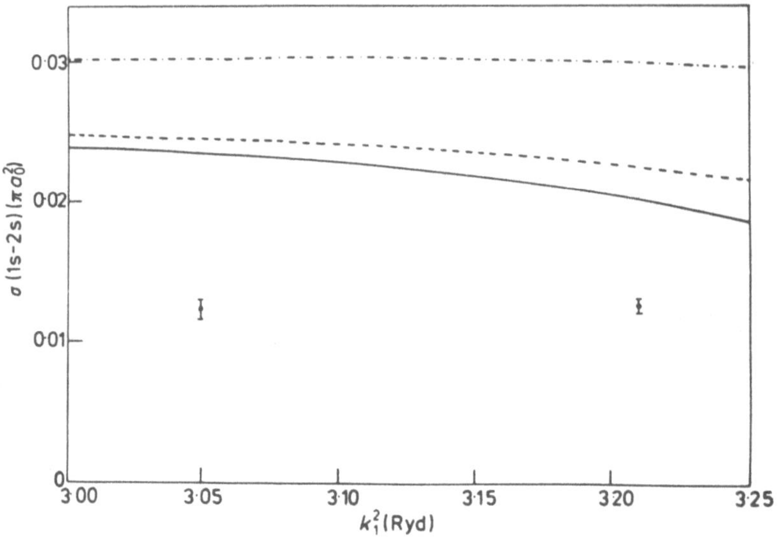

Fig. 6.46 e - He^+ scattering, excitation of the 2s state.
Theory: —— Morgan (1979);
--- Burke and Taylor (1969a);
-·- Burke et al. (1964).
Experiment: ɸ Dolder and Peart (1973).

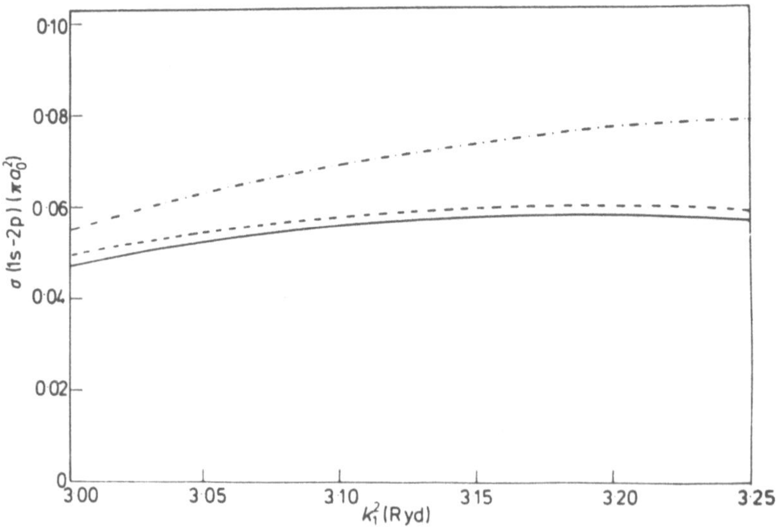

Fig. 6.47 e - He^+ scattering, excitation of the 2p state.
Theory: —— Morgan (1979);
--- Burke and Taylor (1969a);
-·- Burke et al. (1964).

Defining

$$X = \frac{E_0}{\Delta E} \qquad (5.6)$$

where E_0 is the incident electron energy, one can see that the (n,n') resonance occurs at X(n,n') equal to

$$X = \frac{4}{3}\left[1 - \frac{1}{n^2} - \frac{(Z-1)^2}{Z^2}\frac{1}{n'^2}\right]. \qquad (5.7)$$

In the limit $Z \to \infty$, this gives

$$X(3.3) = \frac{28}{27} = 1.037 \ldots$$

which is very near the n = 2 threshold X = 1. Detailed calculations were performed in the "3 states + correlation terms" close coupling approximation for C^{5+} and Ne^{9+}. The results are shown in Figures 6.48 and 6.49 where $Z^2\Omega$ is plotted against $X = E_0/\Delta E$.

Two groups, Kingston and Tayal (1983) and Pradhan et al. (1981), have recently reported results of helium-like ion excitation cross section calculations: in this work the rate coefficients and related quantities were also computed for the purpose of plasma diagnostics.

The effect of dielectronic recombination on resonances was first studied by Presnyakov and Urnov (1975) for the lithium-like ion O^{5+} and recently by Pradhan (1981) for the helium-like ions O^{6+} and Fe^{24+}. In Figure 6.50 the results of Pradhan's work are shown.

D. Intermediate and High-Energy Region

At high energy far away from the excitation threshold the asymptotic form of Ω can be obtained by means of the Coulomb-Born approximation. For optically allowed transitions the collision strengths have the asymptotic form:

$$\Omega(i,f)_{X\to\infty} \sim d\ln 4X, \quad \Delta\ell = 1, \quad \Delta s = 0 \qquad (5.8)$$

where $X = E_0/\Delta E$, E_0 is the energy of incident electron and the slope d is directly proportional to the optical oscillator strength. For ℓ-forbidden transitions, the collision strengths have the asymptotic form:

$$\Omega(i,f)_{X\to\infty} \sim \text{const}; \quad \Delta\ell \neq 1, \quad \Delta s = 0. \qquad (5.9)$$

This constant value may be obtained from plane wave Born approximation calculations. Finally, for spin-forbidden transitions, the collision strengths have the asymptotic form:

$$\Omega(i,f)_{X\to\infty} \sim X^{-2}, \quad \Delta s \neq 0. \qquad (5.10)$$

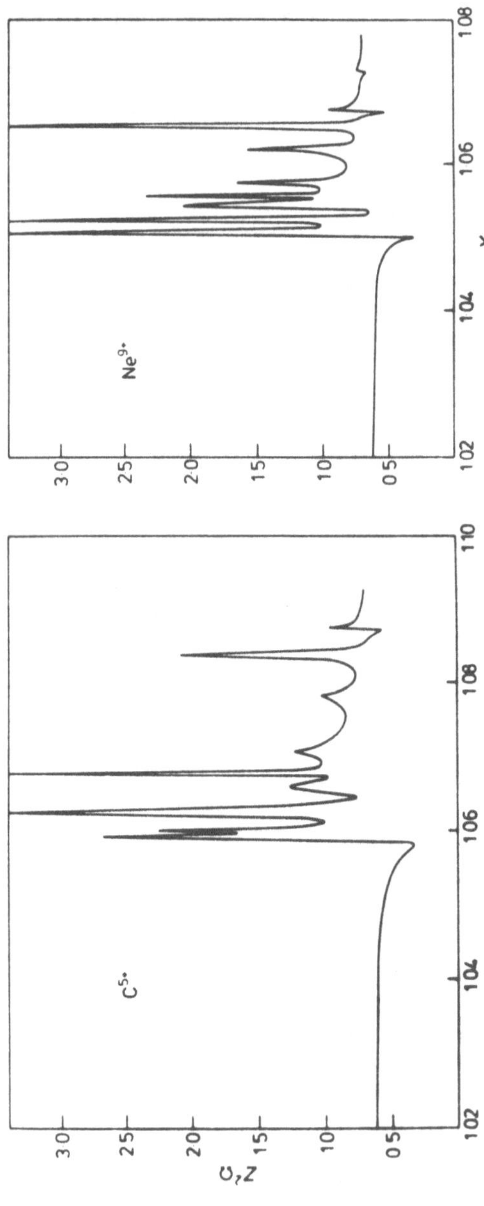

Fig. 6.48 Collision strength for the 1s - 2s transition in C^{5+} and Ne^{9+}.

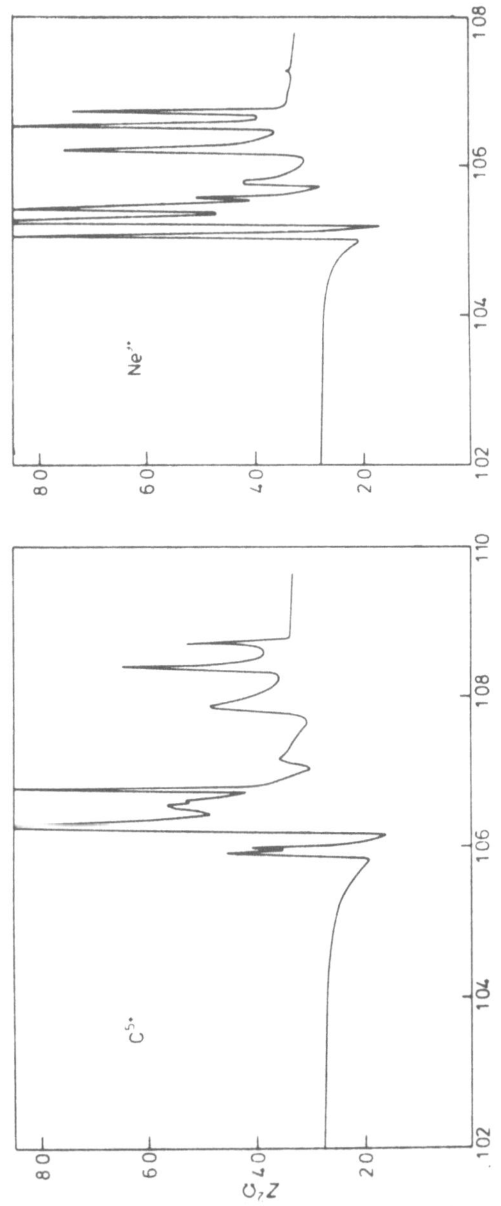

Fig. 6.49 Collision strength for the 1s - 2p transition in Ne^{9+} and C^{5+}.

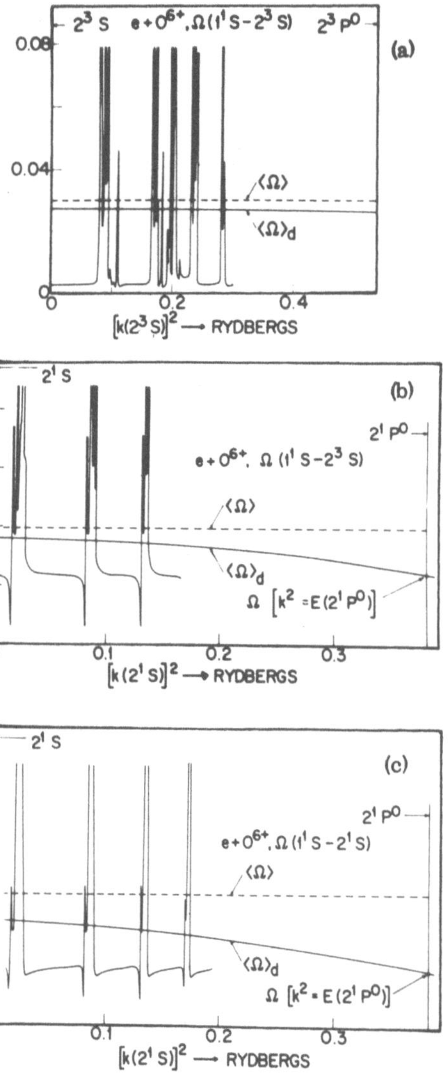

Fig. 6.50 Collision strengths for electron impact excitation of O^{6+}.
a) Transition $1^1S - 2^3S$, energy range $E(2^3S)-E(2^3P^0)$;
b) Transition $1^1S - 2^3S$, energy range $E(2^1S)-E(2^1P^0)$;
c) Transition $1^1S - 2^1S$, energy range $E(2^1S)-E(2^1P^0)$;
$\langle\Omega\rangle$ Collision strength averaged over resonances;
$\langle\Omega\rangle_d$ Averaged collision strength allowing for radiative decay.

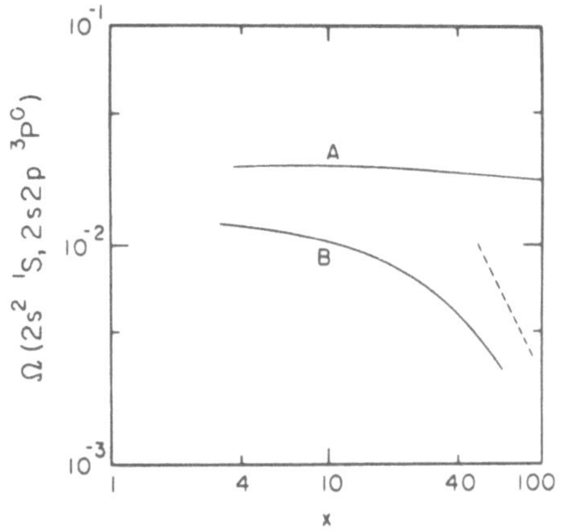

Fig. 6.51 Collision strength for the $(2s)^2\,{}^1S$ - $(2s2p)^3P$ transition in Fe^{22+}. A) close coupling calculation with the mixing of the $2s2p^3P$ and $2s2p^1P$ states included; B) close coupling calculation without mixing (Henry, 1981).

It should be noted, however, that the asymptotic behavior given by (5.8) - (5.10) can be expected only in the case of pure L - S coupling of the target electrons. This behavior can be drastically modified when intermediate coupling plays a significant role due to relativistic mixing of the target states. An example of such a modification is the spin-forbidden transition $(2s)^2\ {}^1S - (2s2p)^3P$ in Fe^{22+} given by Henry (1981). There are two 6-state close coupling calculations quoted by Henry. The following states were used:

$$(2s)^2\ {}^1S,\ 2s2p({}^3P,{}^1P)\ \text{and}\ (2p)^2({}^3P,{}^1D,{}^1S).$$

The spin-orbit interaction mixes the $2s2p^3P$ and $2s2p^1P$ states. One of the quoted calculations neglects the mixing while in the other calculation the mixing is taken into account. The results are compared in Figure 6.51. Curve B corresponds to the first calculation and curve A corresponds to the second.

7
Ionization

7.1 THRESHOLD BEHAVIOR OF THE IONIZATION CROSS SECTION

The behavior of the ionization cross section near threshold was first studied by Wannier (1953). The total energy of both electrons (the incident and the ejected) near threshold is low and the energy dependence of the cross section is determined mainly by the motion of the electron far away from the nucleus. In this case the motion of the electrons can be treated using classical mechanics. This was done by Wannier (1953) and later by Vinkaln and Gailitis (1967).

The simplest situation is when the total energy is precisely zero. Then both electrons move in opposite directions in order to minimize their mutual Coulomb repulsion. Moreover, both electrons are at each moment of time the same distance from the nucleus so that $r_1 = r_2$. Only in this case can they simultaneously overcome the attraction by the atomic nucleus. If r_1 differs significantly from r_2, let us say $r_1 > r_2$, then the kinetic energy of the first electron would increase and that of the second decrease. As a result the second electron would not be able to overcome the nuclear attraction and would remain trapped. There would be no ionization. The classical equations of motion

$$\frac{d^2\mathbf{r}_1}{dt^2} = -\frac{Z\mathbf{r}_1}{r_1^3} + \frac{(\mathbf{r}_1 - \mathbf{r}_2)}{|\mathbf{r}_1 - \mathbf{r}_2|^3}$$

$$\frac{d^2\mathbf{r}_2}{dt^2} = -\frac{Z\mathbf{r}_2}{r_2^3} + \frac{(\mathbf{r}_2 - \mathbf{r}_1)}{|\mathbf{r}_2 - \mathbf{r}_1|^3} \tag{1.1}$$

(in which the electronic charge and mass are considered as unity) in the case of zero total energy can be easily solved. Denoting

$$\mathbf{r}_1 = -\mathbf{r}_2 = \mathbf{r}, \tag{1.2}$$

we get

$$\frac{d^2 r}{dt^2} = -\frac{Z - 1/4}{r^2}, \tag{1.3}$$

The solution of (1.3) is

$$r = \frac{9}{2}(Z - \frac{1}{4})^{1/3} t^{2/3} \tag{1.4}$$

if we adopt the initial condition r = 0 at t = 0 and put the energy equal to zero.

If the total energy is not zero then ionization can proceed even if $\mathbf{r}_1$ is not exactly equal to $-\mathbf{r}_2$. Let us put

$$\begin{aligned} \mathbf{r}_1 &= \mathbf{r} + \Delta\mathbf{r} + \delta\mathbf{r} \\ \mathbf{r}_2 &= -\mathbf{r} + \Delta\mathbf{r} + \delta\mathbf{r} \end{aligned} \tag{1.5}$$

where $\Delta\mathbf{r}$ is the component of $1/2(\mathbf{r}_1 + \mathbf{r}_2)$ along $\mathbf{r}$, and $\delta\mathbf{r}$ is the component of $1/2(\mathbf{r}_1 + \mathbf{r}_2)$ normal to $\mathbf{r}$. Inserting (1.5) into (1.1) we get an equation for $\Delta\mathbf{r}$ and $\delta\mathbf{r}$. If the total energy is small, then it is obvious that only trajectories with small $\frac{|\Delta\mathbf{r}|}{r}$ and $\frac{|\delta\mathbf{r}|}{r}$ lead to ionization. Under the condition $\frac{|\Delta\mathbf{r}|}{r} \ll 1$, $\frac{|\delta\mathbf{r}|}{r} \ll 1$ the equations for $\Delta\mathbf{r}$ and $\delta\mathbf{r}$ reduce to

$$\frac{d^2\Delta\mathbf{r}}{dt^2} = 2Z\frac{\Delta\mathbf{r}}{r^3}, \quad \frac{d^2\delta\mathbf{r}}{dt^2} = -Z\frac{\delta\mathbf{r}}{r^3}. \tag{1.6}$$

In the region Er << Z the relation (1.4) can be used and (1.6) is transformed to

$$\frac{d^2\Delta r}{dt^2} = \frac{16Z}{9(4Z-1)}\frac{\Delta r}{t^2}; \quad \frac{d^2\delta r}{dt^2} = -\frac{8Z}{9(4Z-1)}\frac{\delta r}{t^2}. \tag{1.7}$$

The solution of (1.7) is

$$\Delta r = r(C_1 r^{-1/4-\mu/2} + C_2 r^{-1/4+\mu/2}) \tag{1.8}$$

$$\delta r = r(C_3 r^{-1/4-\nu/2} + C_4 r^{-1/4+\nu/2}) \tag{1.9}$$

where

$$\mu = \frac{1}{2}\sqrt{\frac{100Z-9}{4Z-1}}, \quad \nu = \frac{1}{2}\sqrt{\frac{4Z-9}{4Z+1}} \tag{1.10}$$

[in (1.8) and (1.9) the relation (1.4) is used to express t in terms of r]. It can be shown that near the threshold the total energy of the electron E depends only on the value C_2. Moreover, ionization with a given total energy E can occur if C_2 lies within a certain interval ΔC_2. The length of the interval ΔC_2 is related closely to the cross section ionization. To understand the nature of this relation let us consider what happens near the nucleus. In this region the motion is governed by quantum-mechanical laws. As a result, the electrons enter the outside classical region having a certain probability distribution for the values C_1, C_2, C_3 and C_4. At low energy this distribution is smooth and energy independent. It is thus reasonable to assume a constant probability density for a small interval ΔC_2. Then the probability and hence the cross section will be proportional to the length of the interval ΔC_2. It follows that the problem of determining the energy dependence of the ionization cross section reduces to the problem of determining the energy dependence of a small interval ΔC_2. Here again classical mechanics can help. The crucial point is the scaling relation for the Coulomb field: if $r = f(t)$ is a solution of (1.1), then $r' = B^{-1}f(B^{3/2}t)$ will also be a solution where B is an arbitrary constant. If r corresponds to the energy E, then r' will correspond to $E' = BE$. Using this scaling relation it can be deduced that the length of the interval ΔC_2 leading to ionization should vary with energy as $E^{-1/4+\mu/2}$.

Taking into account that the ionization cross section σ is proportional to ΔC_2 and that the total energy E is equal to $E_0 - I$ where I is the ionization potential and E_0 is the energy of the incident electron, we can write

$$\sigma = \text{const}(E_0 - I)^{-1/4 + \mu/2} \qquad (1.11)$$

which is the Wannier result. For $Z = 1$, $\mu/2 - 1/4 = 1.127$.

The same result was obtained by Peterkop (1971) and Rau (1971) using the semiclassical approximation in quantum mechanics.

Moreover, in the quantum-mechanical treatment it is necessary to consider an additional factor: the total spin and its influence on the threshold law through the symmetry properties of the wave function (Green and Rau, 1982).

The Wannier law was verified by experiments for H in the interval 0 - 0.5 eV (McGowan and Clarke, 1968) and for He in the interval 0 - 2 eV (Krige, 1968). The difference between the threshold behavior in the triplet and singlet states can be observed in the ionization of polarized atoms by polarized electrons. Such experiments are now in progress (Kleinpoppen et al., 1980; Lubel, 1980).

7.2 THE DIFFERENTIAL CROSS SECTIONS: GENERAL RELATIONS

The most detailed characteristic of the ionization process is the triple differential cross section

$$d^3\sigma = \frac{k}{k_o} |A|^2 k'^2 dk' d\Omega d\Omega' . \tag{2.1}$$

Here k_o is the momentum of the incident electron, k' is the momentum of the slow electron moving in a solid angle $d\Omega'$, and k is the momentum of the fast electron moving in a solid angle $d\Omega$. By energy conservation

$$\frac{1}{2}(k'^2 + k^2) = \frac{1}{2} k_o^2 + I. \tag{2.2}$$

If k' and k differ substantially from each other then exchange becomes unimportant and one can identify the slow electron as ejected from the atom. In an experimental study of $d^3\sigma$ the kinematics is fully determined by measuring the energy and momenta of all three electrons involved and by detecting the two final electrons in coincidence. This is often called in the literature an (e, 2e) experiment. Such an experiment is very difficult. Nevertheless there is a growing activity in this field. The reason is that (e, 2e) experiments are very informative. Being performed with fast incident electrons they give in a most direct way important information about the atomic wave function in the initial state. In a simplified version of the Born approximation when the incident electron as well as both outgoing electrons are described by plane waves (this is often called in the literature the "plane wave Born approximation"), the amplitude A is proportional to the Fourier transform of the initial wave function (see Section 7.3). In more accurate and sophisticated approximations the relation between A and the initial wave function is somewhat obscured but still can be revealed (Giardini-Guidoni et al., 1980). At low energy (e, 2e) experiments lead to an insight in the mechanism of the ionization process (Ehrhardt et al., 1980; see also Section 7.5A).

By integrating (2.1) one can form various double differential cross sections. One of them is

$$d^2\sigma = \frac{k}{k_o} k'^2 dk' d\Omega \int |A|^2 d\Omega' . \tag{2.3}$$

This expression can be rewritten in the form

$$\frac{d^2\sigma}{d\varepsilon d\Omega} = \frac{kk'}{k_o} \int |A|^2 d\Omega' \tag{2.4}$$

where $\varepsilon = k'^2/2$ is the energy of the slow moving electron. It follows from (2.2) that

$$\varepsilon = k_o^2/2 - k^2/2 - I.$$

Another double differential cross section is

$$\frac{d^2\sigma}{d\varepsilon d\Omega'} = \frac{kk'}{k_o} \int |A|^2 d\Omega \tag{2.5}$$

which is equivalent to

$$d^2\sigma = \frac{k}{k_o} k'^2 dk' d\Omega' \int |A|^2 d\Omega. \tag{2.6}$$

The third form of double differential cross section is

$$d^2\sigma = \frac{1}{k_o} d\Omega d\Omega' \int_o^{k'_{max}} kk'^2 |A|^2 dk'. \tag{2.7}$$

Integrating once more, we get the differential cross section $d\sigma/d\varepsilon$ which can be classified as the energy loss cross section (by the incident electron). It follows from (2.4) or (2.5) that

$$\frac{d\sigma}{d\varepsilon} = \frac{kk'}{k_o} \int |A|^2 d\Omega d\Omega'. \tag{2.8}$$

7.3 IONIZATION THROUGH EXCITATION OF AUTO-IONIZING STATES. POSTCOLLISION INTERACTION

Looking at the ionization process from the point of view of its mechanism, we can distinguish between direct ionization and resonance ionization through excitation of auto-ionizing states of the target. This leads to the following decomposition of $d^3\sigma$ near the resonance

$$\frac{d^3\sigma}{dk' d\Omega d\Omega'} = \left(\frac{d^3\sigma}{dk' d\Omega d\Omega'}\right)_o + \frac{ax + b}{1 + x^2} \tag{3.1}$$

where

$$x = \frac{2(E_o - E)}{\Gamma},$$

a and b are some functions of all the angular variables involved, and $(d^3\sigma/dk' d\Omega d\Omega')_o$ is the background ionization cross section.

Theoretical calculations of ionization through excitation of auto-ionizing states in noble gases by fast electrons and comparison with experimental data were recently reported by Balashov et al. (1980). At an incident electron energy near the threshold for excitation of an auto-ionizing state an interesting phenomenon occurs which is called the postcollision interaction. Consider the excitation of an auto-ionizing state

$$e + A \rightarrow A^{**} + e.$$

The state A^{**} decays by ejecting an Auger electron

$$A^{**} \rightarrow A^+ + e'$$

(which is marked by a dash). If the energy of the incident electron is much above the threshold then the atom A^{**} will decay long after the electron e went away. But if the energy of the incident electron is near threshold then it is impossible to divide the process into two independent stages.

When the electron e is rather slow it interacts with the atom in the auto-ionizing state A^{**} during all stages of its evolution. This interaction leads to a redistribution of the energy between the scattered electron e and the ejected electron e'. One consequence of this interaction is an apparent shift of the threshold energy. It was first reported by Hicks et al. (1974) (see also Section 7.5B for an example). Another consequence is a peculiar structure in the excitation curves (Smith et al., 1974; Heideman et al., 1974). Theoretical aspects of this phenomenon are considered by Ostrovsky (1977), Morgenstern et al. (1977), Amusia et al. (1980), Read and Comer (1980) and Heideman (1980).

7.4 HYDROGEN

A. Triple Differential Cross Section

The ionization amplitude in the Born approximation is given by the expression

$$A = \frac{2}{q^2} \int \psi_-^*(\mathbf{k}', \mathbf{r}) e^{i\mathbf{q}\cdot\mathbf{r}} \psi_o d\mathbf{r} \tag{4.1}$$

where $\mathbf{q}$ is the momentum transfer $\mathbf{q} = \mathbf{k}_0 - \mathbf{k}$

$$qdq = k_o k \sin\theta d\theta,$$

ψ_o is the wave function of the initial state of the target and ψ_- is the final state wave function. It belongs to the continuum spectrum and has the asymptotic form like "plane wave + ingoing spherical wave" (rather than "outgoing spherical wave"). The reason for this is discussed in scattering theory text books (e.g., Mott and Massey). If the ejected electron is fast enough then ψ_- in (4.1) can be replaced by a plane wave $e^{i\mathbf{k}'\cdot\mathbf{r}}$ which leads to the expression

$$A = \frac{2}{q^2} \int e^{i(\mathbf{k}_0 - \mathbf{k} - \mathbf{k}')\mathbf{r}} \psi_o d\mathbf{r}.$$

In this approximation A is simply proportional to the Fourier transform of ψ_o. Let us return to (4.1). The integral can be evaluated analytically in closed form (Mott and Massey, 1965). The cross section is equal to

$$d^3\sigma = \frac{8\pi}{q^4} \left| \int \psi_-^*(\mathbf{k}', \mathbf{r}) e^{i\mathbf{q}\cdot\mathbf{r}} \psi_o d\mathbf{r} \right|^2 k'^2 dk' d\Omega' qdq. \tag{4.2}$$

This expression does not take into account exchange. To account for exchange the Ochkur approximation can be used which leads to

$$d^3\sigma = 8\pi\left(\frac{1}{k_0^2} \pm \frac{1}{q^2}\right)^2 \left|\int \psi_-^*(\mathbf{k}',\mathbf{r}) e^{i\mathbf{q}\cdot\mathbf{r}} \psi_0 d\mathbf{r}\right|^2 k'^2 dk' d\Omega' q dq. \quad (4.3)$$

Looking at expression (4.1) one can note that the slow and fast electrons after the ionization are treated quite differently. While the slow electron is described by a Coulomb wave function ψ_-, the fast electron is described by a plane wave. This is because the slow electron feels the nuclear Coulomb field while the fast electron feels the Coulomb field screened by the slow electron.

However, in the case where $k' \simeq k$ the above argument fails and it would be more appropriate to treat the two electrons on an equal footing. A short review of various approximations in the theory of (e, 2e) experiments at high energy is given by McCarthy (1980). The results of (e, 2e) experiments for atomic hydrogen and comparison with various calculations are reviewed by Weigold (1980).

B. Double Differential Cross Section. Energy Loss Cross Section

Integrating (4.2) over $d\Omega'$, we get $\frac{d^2\sigma}{dqd\varepsilon}$ in the form

$$\frac{d^2\sigma}{dqd\varepsilon} = 2\pi \frac{2^{10}}{k_0^2 q} \frac{[q^2 + (1+k'^2)/3]\exp[-\frac{2}{k'}\arctan(2k'/q^2 - k'^2 - 1)]}{[(q+k')^2+1]^3[(q-k')^2+1]^3[1-\exp(-2\pi/k')]} . \quad (4.4)$$

When $q \gg 1$ this expression simplifies:

$$\frac{d^2\sigma}{dqd\varepsilon} = \frac{2^6 q}{3k_0^2 k'^5} \frac{1}{[1+(q-k')^2]^3} . \quad (4.5)$$

We can note that this expression has a sharp maximum at $q \simeq k'$ which means that almost all the momentum transfer is carried out by the ejected electron as if it were free.

Integrating (4.5) over q, we have the energy loss cross section

$$\frac{d\sigma}{d\varepsilon} \simeq \frac{2\pi}{k_0^2 \varepsilon} \quad (4.6)$$

which is just the result we can derive from the classical Rutherford formula. If exchange is taken into account then the energy loss spin averaged cross section for $q \gg 1$ is equal to

$$\frac{d\sigma}{d\varepsilon} \simeq \frac{2\pi}{k_0^2}\left[\frac{1}{\varepsilon^2} + \frac{1}{(k_0^2/2-\varepsilon)^2} - \frac{1}{\varepsilon(k_0^2/2-\varepsilon)}\right]. \quad (4.7)$$

The shape of the curve $d\sigma/d\varepsilon$ against ε is symmetrical with respect to the point $x = \varepsilon - k_0^2/4$.

For a small momentum transfer $q \ll 1$ the binding of the atomic electrons becomes important and the expressions (4.6) - (4.7) are inadequate. However, the symmetrical shape of $d\sigma/d\varepsilon$ due to exchange still remains.

C. Total Ionization Cross Section

The total cross section is defined by

$$\sigma = \int_0^{1/2(k_0^2/2 - E)} \frac{d\sigma}{d\varepsilon}\, d\varepsilon \tag{4.8}$$

(the upper limit in the integral corresponds to $k' = k$). To calculate σ for fast incident electrons, the Born approximation with the exchange correction (4.3) can be used. However, for the energy region near threshold the Born approximation becomes inaccurate because both electrons in the final state are slow and should be treated on the same footing.

The most consistent approach of course would be a numerical solution of a three-body problem with Coulomb interactions, at least in classical mechanics. Results of threshold behavior study suggest that the use of classical mechanics can be quite appropriate. The relevant calculations were undertaken by Abrines and Percival (1966) for the ionization of H by protons and by Ochkur and Bratsev (1967) for the ionization of H by electrons. We will not consider here the details of the program and only reproduce the results obtained by Ochkur and Bratsev (1967) (Figure 7.1).

Plotted in Figure 7.1 are the results of the following calculations:

1. Classical mechanics binary approximation.
2. Quantum mechanics Ochkur approximation.
3. 3-body classical mechanics calculation (4 points with statistical error indicated).
4. Experimental data.

Note the rather good agreement of 3 and 4 near the threshold. However, the most important result of this classical calculation is a scaling law for the ionization cross section from an atomic level n, which follows from the equations of motion. If we plot $n^{-4}\sigma_n$ against n^2E, then for all n there will be one and the same curve.

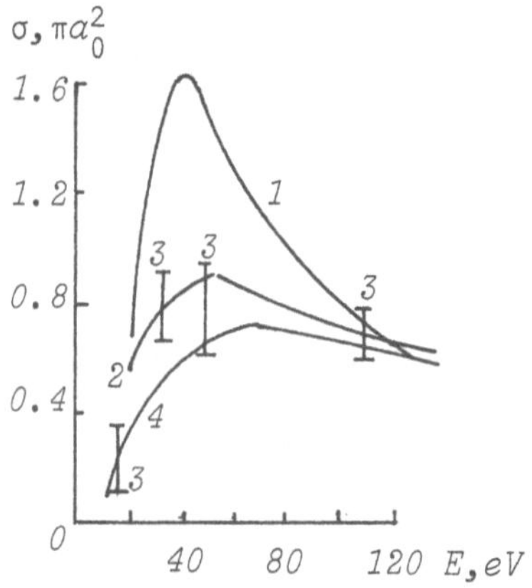

Fig. 7.1 Total cross section for the ionization of H by electrons (Ochkur and Bratsev, 1967).
1. Classical mechanics binary approximation.
2. Quantum mechanics Ochkur approximation.
3. I: 3-body classical mechanics calculations, statistical errors indicated.
4. Experimental data.

7.5 HELIUM

A. Triple Differential Cross Section

At high energy 10^2 - 10^3 eV experimental results together with theoretical calculations were reviewed by Giardini-Guidoni et al. (1980).

We reproduce in Figure 7.2 the squared Fourier transform of the He ground state wave function extracted from experiment together with theoretical values calculated from a Clementi wave function (Clementi and Roetti, 1974). Our main attention will be concentrated on low and intermediate energies. We will consider the results of experiments by Ehrhardt et al. (1980) and their theoretical interpretation.

The notation used in the discussion is explained in Figure 7.3. It is different from those used in this book. $\mathbf{k}_a$, θ_a correspond to our k, θ; k_b, θ_b corresponds to our $\mathbf{k}'$, θ'. The cross section is plotted in a polar diagram as in Figure 7.4 and subsequent figures. The direction $\mathbf{k}_0$ of the incident electron is indicated by the arrow pointing to the center. The direction $\mathbf{k}_a$ of the fast electron is indicated by the arrow going to the upper left from the center. The probability of finding the slow electron at the a given angle θ_b is proportional to the distance between the center and the dots. The

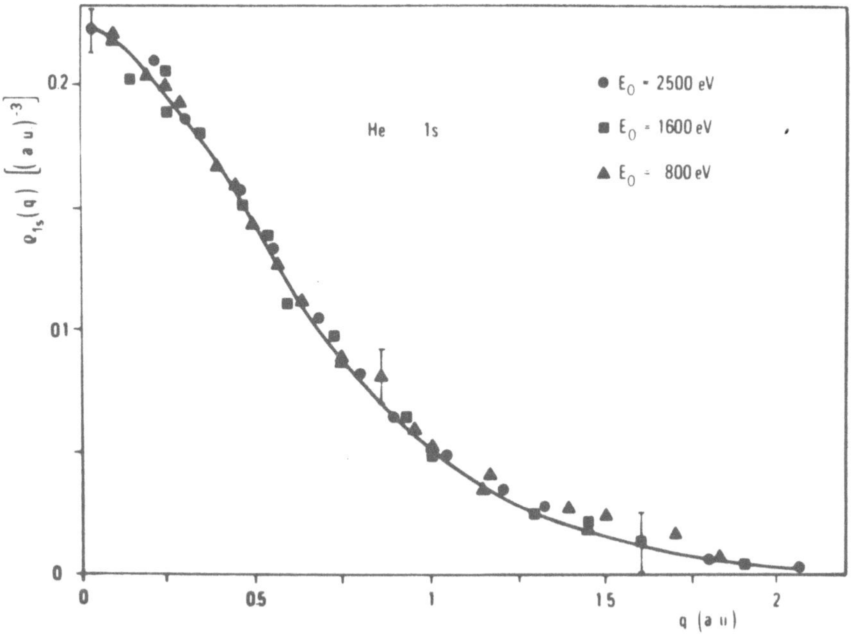

Fig. 7.2 Squared Fourier transform of the He ground state wave function.
Solid line: theory;
Points: experiment with incident electron energies 800, 1600 and 2500 eV (Giardini-Guidoni et al., 1980).

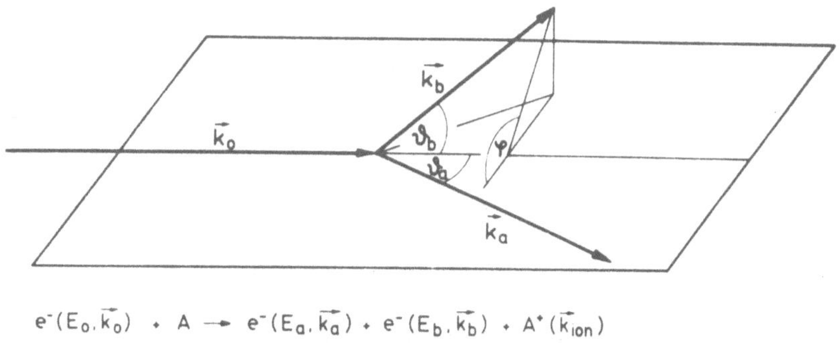

Fig. 7.3 Diagram of kinematics of an ionizing electron collision with an atom (Ehrhardt et al., 1980).

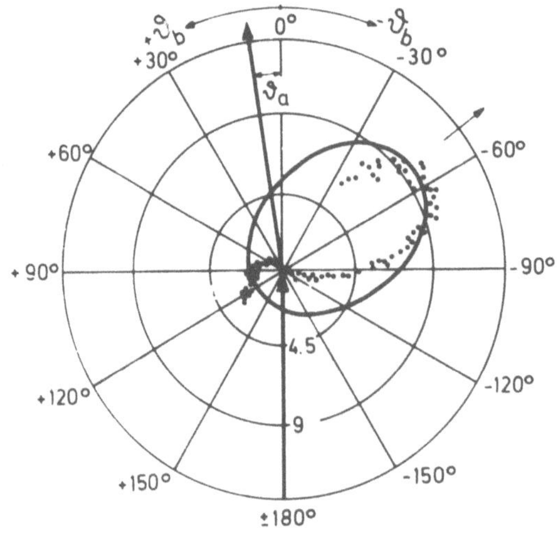

Fig. 7.4 Triple differential cross section polar diagram for He (Ehrhardt et al., 1980).
E_o = 256.5 eV; E_a = 212 eV; E_b = 20 eV; θ_a = 8°.
Solid line: plane wave Born approximation (Glassgold and Jalongo, 1968).

small arrow indicates the direction of momentum transfer $\mathbf{q} = \mathbf{k}_o - \mathbf{k}_a$. At incident energies $\gtrsim 10^2$ eV the angular distribution of the slow electron displays two distinct peaks nearly opposite to each other and sharp minima between them. An example is shown in Figure 7.4 (E_o = 256.5 eV, E_a = 212 eV, E_b = 20 eV and $\theta_a = 8^\circ$) and Figure 7.5 ($\theta_a = 4^\circ$, all other parameters are the same as in Figure 7.4). The forward peak in the direction of **q** is called the binary peak. It corresponds to the situation when almost all **q** is transferred to the ejected electron as if there is a two-body collision. Note that for elastic collisions (without energy loss) between two free electrons, the angle between $\mathbf{k}_a$ and $\mathbf{k}_b$ would be 90°. But actually in our case the collision is inelastic and we can see in Figures 7.4 and 7.5 that the angle is much less than 90°. The second peak is called the recoil peak. It corresponds to a redistribution of the momentum transfer **q** from the atomic electron to the nucleus.

The theoretical predictions are indicated by the solid-line curves. In Figure 7.4 the result of the plane wave Born approximation is indicated (Glassgold and Jalongo, 1968). We expect that this calculation fails to reproduce the recoil peak because the interaction of the slow electron with the nucleus is not taken into account. The theoretical result in Figure 7.5, indicated by the solid-line curve, is obtained using the Born approximation with a Coulomb field wave function for the slow electron (Veldre and Vinkaln,

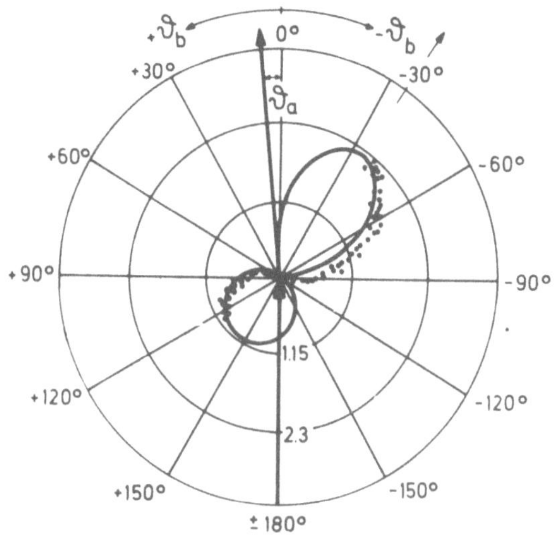

Fig. 7.5 Triple differential cross sections polar diagram for He (Ehrhardt et al., 1980).
E_o = 256.5 eV; E_a = 212 eV, E_b = 20 eV and $\theta_a = 4^\circ$.
Solid line: Born approximation (Veldre and Vinkalin, 1966).

1966). If we look at the angular distribution at lower energy it will be quite different from that shown in Figures 7.4 - 7.5. As an example let us consider the case E_o = 50 eV, E_a = 20 eV, E_b = = 5.5 eV and $\theta_a = 7^\circ$ (Figure 7.6). The shape of the angular distribution is rather complicated. Moreover, the classification of peaks as "binary" and "recoil" becomes meaningless, because there is no correlation with the direction of momentum transfer **q**.

At still lower energies some new features appear in the angular distribution. Consider for example the case E_o = 30.5 eV. In Figure 7.7 the angular distribution is shown for two partitions of the energy: solid circles – $E_b = E_a$ = 3 eV; triangles – E_a = 2 eV, E_b = 4 eV. The angle θ_a in both cases is 120°. We can see that the shape of the angular distribution remains the same. The threshold behavior theories predict such independence. But the shape itself is very far from the strong angular correlation required by the Wannier threshold theory.

There are no theoretical calculations at present which can reproduce the main characteristic features of the angular distributions at low energy.

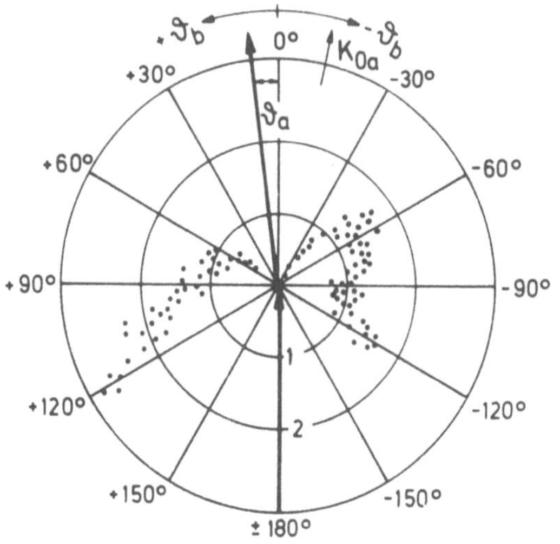

Fig. 7.6 Triple differential cross section polar diagram for He (Ehrhardt et al., 1980).
E_o = 50 eV; E_a = 20 eV; E_b = 5.5 eV; θ_a = 7°.

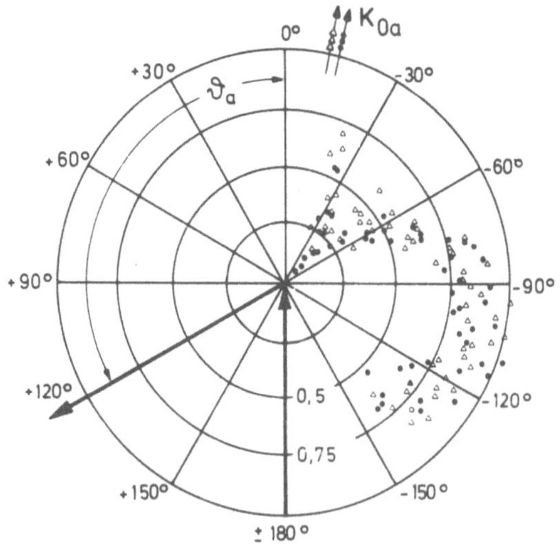

Fig. 7.7 Triple differential cross section polar diagram for He (Ehrhardt et al., 1980).
E_o = 30.5 eV; θ_a = 120°; ● – $E_b = E_a$ = 3 eV; Δ – E_a = 2 eV; E_b = 4 eV.

B. Double Differential Cross Sections

The theoretical and experimental results for $d^2\sigma/d\varepsilon d\Omega$ at high incident electron energy were reviewed by Oda (1973).

For small energy loss $\lesssim$ 5 - 10%, there is good agreement between the Born approximation and experiment. However, for a greater energy loss $\sim$ 50% the agreement is not so good.

The cross section $d^2\sigma/d\varepsilon d\Omega'$ was calculated by Balashov et al. (1973) for various energies of the slow electron as a function of the angle of its motion θ' (Figure 7.8). An interesting feature of the set of curves is the gradual formation of a binary peak with the increasing energy of the slow electron. Finally, some recent experimental results concerning the postcollision interaction in helium will be considered.

Roy (1980) reported measurements of the differential electron energy-loss spectra $d^2\sigma/d\varepsilon d\Omega$ in our notation. All measurements were carried out in the forward direction ($\theta = 2^\circ$), the residual energy of the scattered electron E_s ranging from 1 to about 40 eV.

In Figure 7.9 the energy loss spectra for various residual energies are reproduced. In the upper spectrum the features

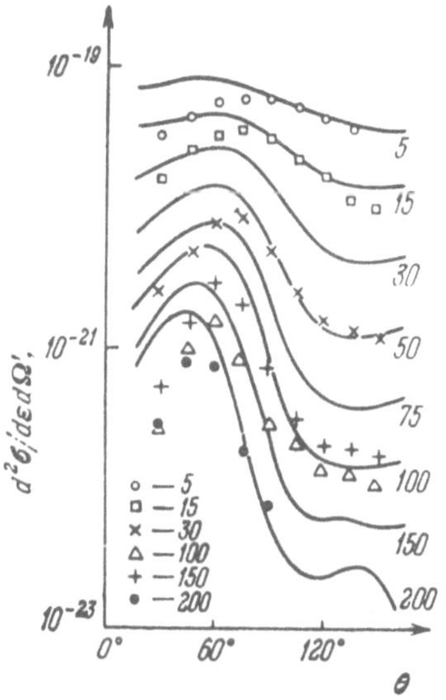

Fig. 7.8 Angular distribution of ejected electrons from He at various energies. The incident electron energy is fixed. E_0 = 500 eV (Balashov et al., 1973); cross section in units $cm^2/eV \cdot rad$; energies in eV.

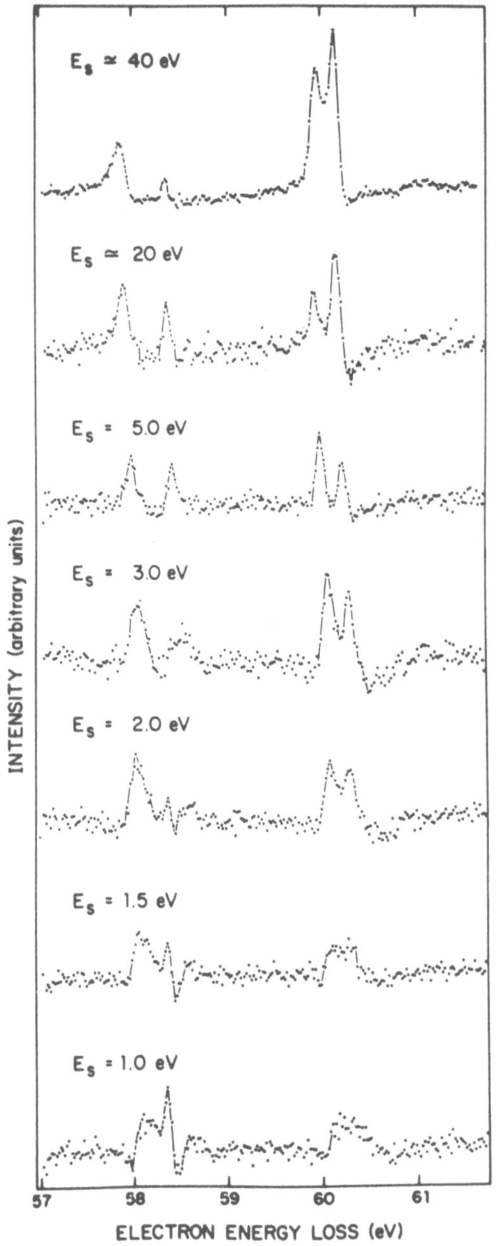

Fig. 7.9 Differential energy-loss spectra in He, measured in the forward direction for various values of the residual scattered electron energy (Roy, 1980).

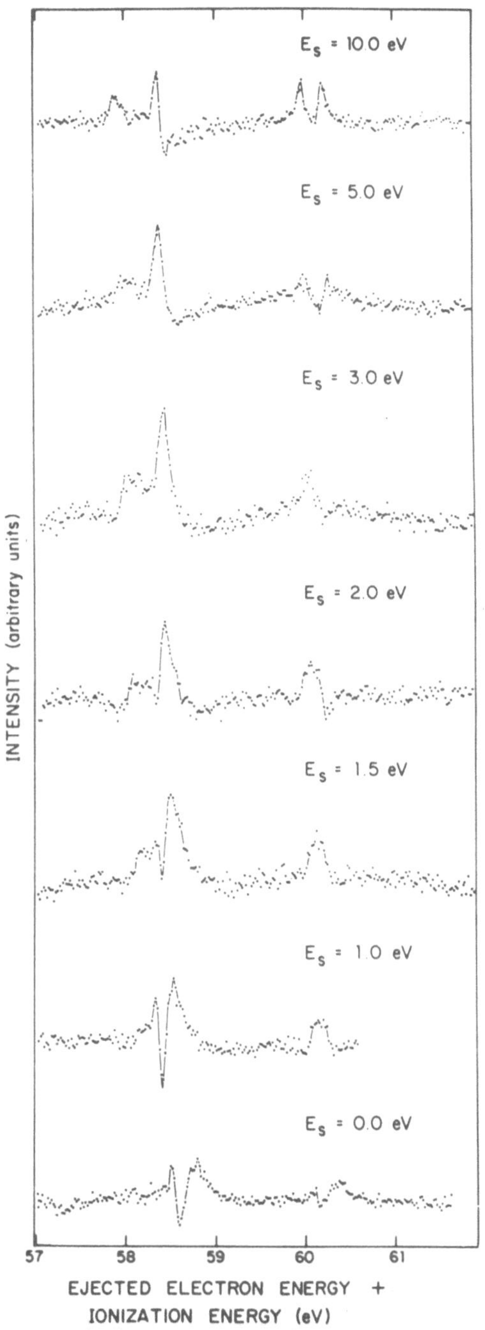

Fig. 7.10 Differential ejected electron spectra (Roy, 1980).

correspond to the four auto-ionizing states $(2s^2)^1S$, $(2s2p)^3P$, $(2p^2)^1D$ and $(2s2p)^1P$ states. In the lower spectra, the shapes, relative magnitudes and energy positions of the features are modified by the postcollision interaction. In Figure 7.10 the ejected electron spectra are reproduced. The energy scale is shifted by an amount corresponding to the ionization energy 24.59 eV. The features around 58 eV in particular exhibit complicated structures.

7.6 TOTAL IONIZATION CROSS SECTIONS FOR VARIOUS ATOMS

A. Born Approximation Calculations

The total cross section is defined by (4.8). It was calculated in the Born approximation by Omidvar et al. (1972) for He, Li, C, N, O, Ne, Na, Mg, Ar, K and Zn, the ejection of s, p and d electrons being taken into account; by McGuire (1971) for all atoms between He and Na; by Peach (1971) for all atoms from Be to Ne and from Al to Ar and for a number of atomic ions; and by Vainstein et al. (1971) for Ba, Sr, Ca and Mg, the contribution of inner shells being taken into account. Some results of the above quoted papers are in Figures 7.11 and 7.12.

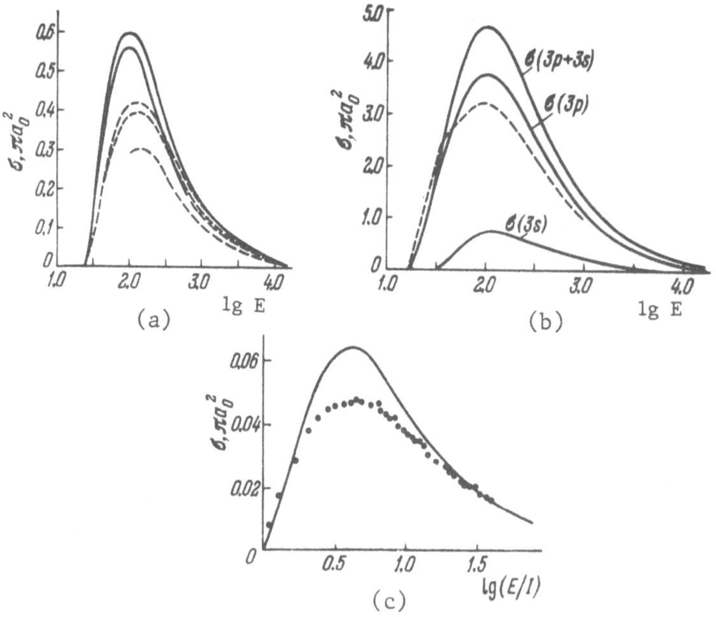

Fig. 7.11 Total ionization cross sections calculated in the Born approximation.
a) He: solid lines, various versions of the Born approximation; dashed lines, experimental data.
b) Ar: solid lines, theory; dashed line, experimental.
c) Li^+: solid line, theory; dots, experimental data.

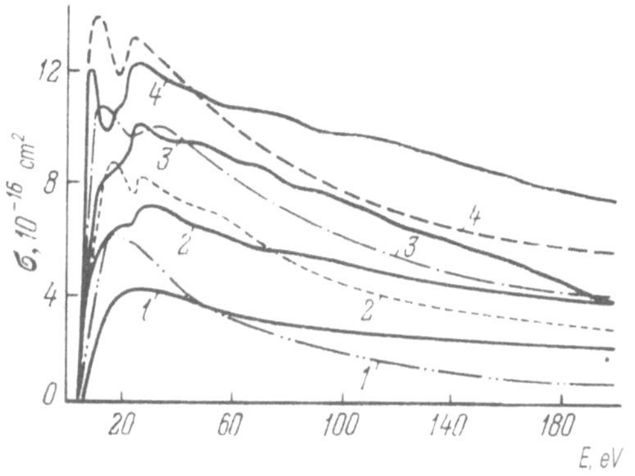

Fig. 7.12 Total ionization cross section for (1) Mg, (2) Ca, (3) Sr and (4) Ba.
Solid line: experimental data; dashed line: theory; contribution of inner shells included.

B. Classical Binary Approximation

The application of the classical binary approximation to the hydrogen atom was considered in Section 5.4. If we try to extend this approach to more complicated atoms, then the first problem will be to find a reasonable value for the mean kinetic energy of an atomic electron which is ejected. This problem was studied by Ochkur. It seems that this quantity cannot be established by purely

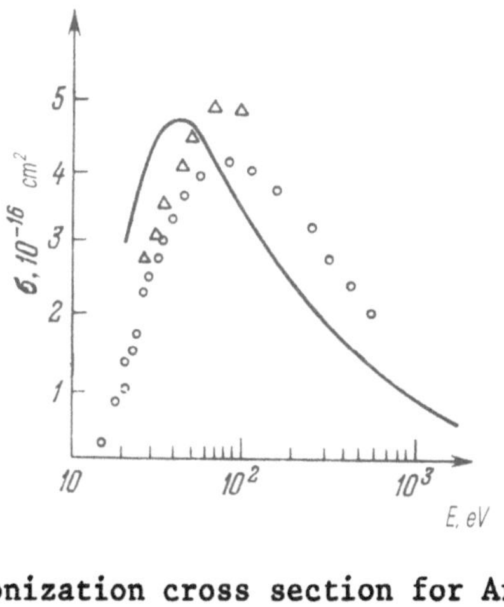

Fig. 7.13 Total ionization cross section for Ar, calculated in the classical binary approximation (Ochkur, 1976).
Solid line: theory; Δ, ○: various experimental data.

theoretical arguments. Rather, it should be considered as an adjustable empirical parameter. For instance, for Ar a good fit to the experimental results can be obtained if we put the mean kinetic energy at 4.6 times the ionization potential (Figure 7.13).

8
Rotational and Vibrational Excitation of Molecules

8.1 EXCITATION OF ROTATION OF TWO ATOM MOLECULES BELOW THE VIBRATIONAL THRESHOLD

In this energy region the excitation of rotation is mainly due to the long range interaction of the electron with the electric multipole moments of the molecule. This was shown first by Gerjuoy and Stein (1955) who took into account only the quadrupole moment. Later their results were extended by Takayanagi and Itikawa (1970).

Assuming that the incident electron velocity is much larger than the mean velocity of the nuclei in the rotationally relevant excited states, we can use the adiabatic approximation. First calculate the elastic scattering amplitude of an electron by a molecule with fixed nuclei $A(\mathbf{k},\mathbf{k}_0,\mathbf{R}^0)$. The transition amplitude $\langle j'm'|A|jm\rangle$ is then related to $A(\mathbf{k},\mathbf{k}_0,\mathbf{R}^0)$ by

$$\langle j'm'|A|jm\rangle = \int Y^*_{j'm'}(\mathbf{R}^0)A(\mathbf{k},\mathbf{k}_0,\mathbf{R}^0)Y_{jm}(\mathbf{R}^0)d\Omega. \qquad (1.1)$$

Here $\mathbf{R}^0$ is the unit vector along the molecular axis. The integration in (1.1) is over all orientations of $\mathbf{R}^0$. The quantum numbers m, m' refer to a certain fixed quantization axis. Usually one is interested in the cross section summed over final m' and averaged over initial m. This cross section in fact will not depend on the particular choice of direction of the quantization axis.

Note that because of the large difference between the electronic and nuclear mass, it can happen that the electronic energy will be near the threshold energy for rotational excitation while the electronic velocity still substantially exceeds the nuclear velocity.

It is advantageous to expand $A(\mathbf{k},\mathbf{k}_o,\mathbf{R}^0)$ in spherical harmonics

$$A(\mathbf{k},\mathbf{k}_o,\mathbf{R}^0) = \sum_\lambda \sqrt{\frac{4\pi}{2\lambda+1}}\, A_\lambda(\mathbf{k},\mathbf{k}_o) Y_{\lambda o}(\mathbf{R}^0). \tag{1.2}$$

Inserting (1.2) in (1.1) and using the relation

$$\int Y_{\ell_1 m_1} Y_{\ell_2 m_2} Y_{\ell_3 m_3} d\Omega =$$

$$= [(2\ell_1+1)(2\ell_2+1)(2\ell_3+1)/4\pi]^{\frac{1}{2}} \begin{pmatrix} \ell_1 & \ell_2 & \ell_3 \\ 0 & 0 & 0 \end{pmatrix} \begin{pmatrix} \ell_1 & \ell_2 & \ell_3 \\ m_1 & m_2 & m_3 \end{pmatrix} \tag{1.3}$$

where $\begin{pmatrix} \ell_1 & \ell_2 & \ell_3 \\ m_1 & m_2 & m_3 \end{pmatrix}$ is the Wigner 3j-symbol, we have

$$\langle j'm'|A|jm\rangle = \sum_\lambda \sqrt{(2j'+1)(2j+1)} \begin{pmatrix} j' & \lambda & j \\ 0 & 0 & 0 \end{pmatrix} \begin{pmatrix} j' & \lambda & j \\ m' & 0 & m \end{pmatrix} A_\lambda(\mathbf{k},\mathbf{k}_o). \tag{1.4}$$

The differential cross section for a transition $jm \to j'm'$ is equal to

$$\frac{d\sigma_{j'm'jm}}{d\Omega} = \frac{k}{k_o} |\langle j'm'|A|jm\rangle|^2.$$

Summing over m' and averaging over m, we obtain the cross section $d\sigma_{j'j}/d\Omega$:

$$\frac{d\sigma_{j'j}}{d\Omega} = \frac{k}{k_o} \sum_{m'm} \frac{1}{2j+1} |\langle j'm'|A|jm\rangle|^2.$$

Inserting (1.4) and using the relation

$$\sum_{m'm} \begin{pmatrix} j' & \lambda & j \\ m' & 0 & m \end{pmatrix} \begin{pmatrix} j' & \lambda' & j \\ m' & 0 & m \end{pmatrix} = \frac{\sigma_{\lambda\lambda'}}{2\lambda+1},$$

we obtain

$$\frac{d\sigma_{j'j}}{d\Omega} = \frac{k}{k_o} \sum_\lambda \frac{2j'+1}{2\lambda+1} \begin{pmatrix} j' & \lambda & j \\ 0 & 0 & 0 \end{pmatrix}^2 |A_\lambda(\mathbf{k},\mathbf{k}_o)|^2. \tag{1.5}$$

Integrating over the angular variables and using the relation

$$q dq = k\, k_o \sin\theta d\theta$$

where $\mathbf{q} = \mathbf{k}_o - \mathbf{k}$ is the momentum transfer, we obtain for the cross section

$$\sigma_{j'j} = \frac{2\pi}{k_o^2} \sum_\lambda \frac{2j'+1}{2\lambda+1} \begin{pmatrix} j' & \lambda & j \\ 0 & 0 & 0 \end{pmatrix}^2 \int_{k-k_o}^{k+k_o} |A_\lambda|^2 q dq. \tag{1.6}$$

Let us represent (1.5) in another form. If we put $j = 0$ in (1.5) and use the relation

$$\begin{pmatrix} j' & \lambda & 0 \\ 0 & 0 & 0 \end{pmatrix}^2 = \frac{1}{2j'+1} \delta_{\lambda j'},$$

then

$$\frac{d\sigma_{j'0}}{d\Omega} = \frac{k}{k_o}\frac{1}{2j'+1}\,|A_{j'}(\mathbf{k},\mathbf{k}_o)|^2. \tag{1.7}$$

Combining (1.5) and (1.7), we can deduce that

$$\frac{d\sigma_{j'j}}{d\Omega} = \sum_{\lambda}\begin{pmatrix} j' & \lambda & j \\ 0 & 0 & 0\end{pmatrix}^2 \frac{d\sigma_{\lambda 0}}{d\Omega}\,. \tag{1.8}$$

This is a very useful relation because it expresses the cross section of a transition from an excited state $d\sigma_{j'j}/d\Omega$ in terms of cross sections from the ground state. This relation was derived and used by many authors (Chu and Dalgarno, 1975; Warshalovich et al., 1977; Goldflam et al., 1977; Khare, 1977; Shimamura, 1979; Lavrov et al., 1980).

If the energy of the incident electron is far above the threshold then $k \simeq k_o$ and one can sum the differential cross section also over all final momenta j'. The result will represent the total excitation cross section $d\bar{\sigma}/d\Omega$. To perform this summation let us turn to (1.5). Putting $k = k_o$ we have

$$\frac{d\bar{\sigma}}{d\Omega} = \sum_{j'}\frac{d\sigma_{j'j}}{d\Omega} = \sum_{mm'j'}|\langle j'm'|A|jm\rangle|^2\frac{1}{2j+1}$$

$$= \sum_{mm'j'}\frac{1}{2j+1}\langle jm|A^+|j'm'\rangle\langle j'm'|A|jm\rangle$$

$$= \sum_{m}\frac{1}{2j+1}\langle jm|A^+A|jm\rangle$$

$$= \sum_{m}\frac{1}{2j+1}\int Y^*_{\ell m}|A(\mathbf{k},\mathbf{k}_o,\mathbf{R}^0)|^2 Y_{\ell m}\,d\Omega.$$

Using the identity

$$\frac{1}{2j+1}\sum_{m} Y^*_{\ell m}Y_{\ell m} = \frac{1}{4\pi}\,,$$

we have finally

$$\frac{d\bar{\sigma}}{d\Omega} = \frac{1}{4\pi}\int|A(\mathbf{k},\mathbf{k}_o,\mathbf{R}^0)|^2 d\Omega. \tag{1.9}$$

Note that the right hand side of (1.9) is the elastic scattering cross section from a molecule with fixed nuclei, averaged over all directions of the molecular axis.

To determine the amplitude $A(\mathbf{k},\mathbf{k}_o,\mathbf{R}^0)$ we can use the Born approximation. The reason is that for long range multipole potentials the main contribution to the scattering amplitude comes from rather large distances. But at large distances the absolute value of the potential is small and the potential can be treated as a small perturbation.

However, there is an important exception: a polar molecule with a large dipole moment. If the dipole moment exceeds a certain critical value then bound states of an electron in a dipole field become possible. This leads to resonances which cannot be treated in the Born approximation (Fabrikant, 1977).

This approach can also be applied to multi-atom molecules. However, a detailed account involves the quantization theory of tops which is beyond the scope of this book. We will mention only two papers by Itikawa (1971). An interesting point is the unusually large cross sections at low energy in some cases. For instance, the cross section of the first rotational state excitation in H_2O at 10^{-2} eV is of the order 10^{-12} cm^2. Let us now consider in detail the amplitude $A(\mathbf{k},\mathbf{k}_o,\mathbf{R}^0)$ in the Born approximation (Section 1.8)

$$A = -\frac{1}{2\pi}\int e^{i\mathbf{q}\cdot\mathbf{r}}U d\mathbf{r} \tag{1.10}$$

where U is the interaction energy.

Representing U in the form

$$U = \sum_\lambda U_\lambda(r)P_\lambda(\cos\theta), \tag{1.11}$$

and using the expansion of a plane wave (2.25) of Chapter 1, we can reduce the amplitude $A(\mathbf{k},\mathbf{k}_o,\mathbf{R}^0)$ to the form (1.2) where A_λ is related to U_λ by

$$A_\lambda = -2i^\lambda\int_0^\infty dr\, r^2 J_{\lambda+\frac{1}{2}}(qu)U_\lambda(r). \tag{1.12}$$

Let us divide the range of integration in (1.12) into two regions from 0 to a certain value a inside the molecule, and from a to ∞ outside the molecule. In the outside region we put

$$U_\lambda = -Q_\lambda/r^{\lambda+1} \tag{1.13}$$

where Q_λ is some constant parameter (provided $\lambda \neq 0$). In the case of a neutral molecule there is no term like Q_o/r. The potential U_o decreases faster than $1/r$ in the limit $r \to \infty$, and does not contribute to rotational excitation. The inner part does not contribute to rotational excitation for all $\lambda \geqslant 1$ as long as $ka \ll 1$.

Note that the term $Q_\lambda/r^{\lambda+1}$ represents the permanent multipole moment of the molecule. But in addition there are multipole moments, induced by the electric field of a scattered electron. The most important among them is the induced dipole moment. In the case of electron-atom scattering an induced dipole moment leads to a spherically symmetrical additional term in the interaction energy $-\alpha/r^4$. In the case of electron-molecule scattering there will be also an anisotropic part of polarization interaction $-\alpha'/r^4P_2(\cos\theta)$. The angular dependence of this term $P_2(\cos\theta)$ corresponds to a quadrupole momentum, but the radial part does not behave like r^{-3} as we can expect for a pure quadrupole interaction. In fact, we have r^{-4} instead of r^{-3}. So for $\lambda = 2$, we obtain

$$U_2 = -(Q_2/r^3 + \alpha'/r^4). \tag{1.14}$$

Now in the integral over the outside region (a,∞) we can, without introducing large error, replace a by 0 as long as $ka \ll 1$. Taking all these considerations into account we obtain for the most important dipole and quadrupole terms

$$A_1 = 2i\frac{Q_1}{q}; \quad A_2 = -2\left(\frac{Q_2}{3} + \frac{\pi\alpha' q}{16}\right). \tag{1.15}$$

Inserting (1.15) into (1.6) we can calculate the contribution to $\sigma_{j'j}$ due to the dipole and quadrupole terms, $\sigma^{(1)}_{j'j}$ and $\sigma^{(2)}_{j'j}$. Let us consider first $\sigma^{(1)}_{j'j}$

$$\sigma^{(1)}_{j'j} = \frac{8\pi}{3k^2}(2j'+1)\begin{pmatrix} j' & 1 & j \\ 0 & 0 & 0\end{pmatrix}^2 Q_1^2 \ln\frac{k+k'}{k-k'}. \tag{1.16}$$

The Wigner 3 − j symbol $\begin{pmatrix} j' & 1 & j \\ 0 & 0 & 0\end{pmatrix}$ has a nonzero value only when $j' = j \pm 1$. This relation determines the allowed transitions for the dipole interaction; $j' = j + 1$ corresponds to an excitation process while $j' = j - 1$ corresponds to de-excitation. For the case $j' = j + 1$

$$\begin{pmatrix} j' & 1 & j \\ 0 & 0 & 0\end{pmatrix}^2 = \frac{j+1}{2(2j+1)(2j+3)},$$

and

$$\sigma^{(1)}_{j+1,j} = \frac{8\pi}{3k^2} Q_1^2 \frac{j+1}{2(2j+1)} \ln\frac{k-k'}{k+k'}. \tag{1.17}$$

Next, let us consider $\sigma^{(2)}_{j'j}$

$$\sigma^{(2)}_{j'j} = \frac{8\pi}{k^2}\frac{2j'+1}{5}\begin{pmatrix} j' & 2 & j \\ 0 & 0 & 0\end{pmatrix}^2 \int_{k-k'}^{k+k'} dq\, q\left(\frac{Q_2}{3} + \frac{\pi\alpha' q}{16}\right)^2. \tag{1.18}$$

The Wigner 3 − j symbol $\begin{pmatrix} j' & 2 & j \\ 0 & 0 & 0\end{pmatrix}$ has a nonzero value only if $j' = j \pm 2$. This determines the allowed transitions for the quadrupole interaction. For the case $j' = j + 2$

$$\begin{pmatrix} j' & 2 & j \\ 0 & 0 & 0\end{pmatrix} = \frac{3(j+2)(j+1)}{2(2j+5)(2j+3)(2j+1)}$$

and

$$\sigma^{(2)}_{j+2,j} = \frac{12\pi}{5k^2}\frac{(j+1)(j+2)}{(2j+1)(2j+3)} \int_{k-k'}^{k+k'} dq\, q\left(\frac{Q_2}{3} + \frac{\pi\alpha' q}{16}\right)^2. \tag{1.19}$$

Near threshold the term $\pi\alpha' q$ can be neglected. Then

$$\sigma^{(2)}_{j+2,j} - \frac{8\pi}{15} Q_2^2 \frac{(j+1)(j+2)}{(2j+1)(2j+3)} \sqrt{1-\frac{2\Delta E}{k^2}} . \qquad (1.20)$$

Far above the threshold we can approximately replace the lower limit in the integral by zero and the upper limit by 2k. Then

$$\sigma^{(2)}_{j+2} = \frac{12\pi(j+1)(j+2)}{5(2j+1)(2j+3)} \left[\frac{2Q_2^2}{9} + \frac{\pi Q_2 \alpha' k}{9} + \frac{\pi^2 \alpha'^2 k^2}{64}\right]. \qquad (1.21)$$

If instead of a neutral molecule we consider a positive molecular ion, then the Born approximation should be replaced by the Coulomb Born approximation. This means that Coulomb wave functions are used instead of plane waves in (1.10). The calculation of the partial amplitude A_λ is similar to the calculation of atomic ion excitation in the Coulomb–Born approximation (Gailitis, 1963). We reproduce here the results obtained by Bojkova and Objedkov (1968). For the dipole excitation case

$$\sigma^{(1)}_{j'j} = \frac{2\pi Q_1^2}{3\Delta E} \frac{j+1}{2j+1} \sigma_0(E, \frac{E}{\Delta E}). \qquad (1.22)$$

Here σ_0 is the function which was introduced by Gailitis (1963):

$$\sigma_0 = 2\pi^2 \frac{\Delta E}{E} \frac{e^{2\pi\eta}}{(e^{2\pi\eta}-1)(e^{2\pi\eta'}-1)} x_0 \frac{d}{dx_0} |F(i\eta, i\eta', 1, x_0)|^2,$$

$$\eta = \frac{1}{k}, \quad \eta' = \frac{1}{k'}, \quad x_0 = -\frac{4\eta\eta'}{(\eta'-\eta)^2}, \qquad (1.23)$$

where F is the hypergeometric function. Near the threshold, $E \sim \Delta E$ and

$$\sigma^{(1)}_{j'j} = \frac{4\pi}{3\sqrt{3}} \frac{\pi}{k^2} Q_1^2 \frac{j+1}{2j+1} . \qquad (1.24)$$

Note the absence of the term $\sqrt{1 - 2\Delta E/k^2}$ which is present in (1.20). As a result the cross section at threshold becomes very large. Far from the threshold the influence of the Coulomb field becomes negligible. For the quadrupole excitation near threshold, $\Delta E \sim E$ and

$$Q^{(2)}_{j+2,j} = \frac{\pi Q_2^2}{k^2} 0.16 \frac{(j+1)(j+2)}{(2j+1)(2j+3)} . \qquad (1.25)$$

Again the factor $\sqrt{1 - 2\Delta E/k^2}$ does not enter in contrast with (1.20).

8.2 EXCITATION OF VIBRATIONS AND COMBINED ROTATIONAL-VIBRATIONAL TRANSITIONS

If the characteristic time of the interaction between an electron and a molecule is much shorter than the period of vibration, again the adiabatic approximation can be used. The amplitude of a transition from an initial state 0, characterized by vibrational and rotational quantum numbers v_o, j_o, m_o to a given final state n with quantum numbers v, j, m is equal to

$$A_{no} = \int \zeta_n^*(\mathbf{R})A(\mathbf{R})\zeta_o(\mathbf{R})d\mathbf{R}. \tag{2.1}$$

Here $\zeta_n = \phi_v(\mathbf{R})Y_{jm}(\mathbf{R}^0)$ where ϕ_v, the vibrational wave functions, are assumed to be independent of the rotational quantum numbers j, m. $A(\mathbf{R})$ is the elastic scattering amplitude by a molecule with fixed nuclei.

The effective cross section for the transition under consideration is

$$\sigma_{no} = \frac{k_n}{k_o} \int |A_{no}|^2 d\Omega. \tag{2.2}$$

If the energy of the incident electron is far above the vibrational threshold then $k_n \simeq k_o$ and we can determine the total excitation cross section summed over all n [in a manner similar to the derivation of (1.9)]. It is equal to

$$\bar{\sigma} = \int |\phi_o|^2 \sigma(\mathbf{R}) d\mathbf{R} \tag{2.3}$$

where $\sigma(\mathbf{R})$ is the elastic scattering cross section by a molecule with fixed nuclei. According to (2.3) it is averaged over the zero energy vibrations of the nuclei in the molecule.

The interaction between an electron and a molecule can be divided into two regions: long range and short range. In contrast to the case of pure rotational excitation at low energy, both parts are important for vibrational excitation. It is important to realize that the Born approximation is not adequate for the treatment of the short-range part. The reason is that in the Born approximation the scattering amplitude from two atoms in the molecule would be a sum of the scattering amplitudes from each atom separately. But actually multiple scattering of an electron inside the molecule is very important. In this case the scattering amplitudes are by no means additive. To demonstrate the role of multiple scattering in the most simple and explicit way we will apply the zero-range potential model. The interaction of an electron with a molecule is replaced approximately by an interaction with each atom as if they were simple interaction centers. The interaction with an atom is replaced in turn by a boundary condition for the wave function similar to (4.19) of Chapter 1. This boundary condition was formulated for a spinless particle. Now we will generalize it to include spin variables because the real electron-atom interaction depends on the total spin.

We will write the boundary condition in the form

$$\frac{d}{dr}(r\psi)_o = -\left(\frac{1}{a_s}P_s + \frac{1}{a_t}P_t\right)(r\psi)_o \tag{2.4}$$

where a_s and a_t are the scattering lengths in the singlet and triplet states of the system, and P_s and P_t are the corresponding projection operators acting on the spin variables of the wave function. They can be expressed in terms of Pauli spin matrices by the well-known relations

$$P_s = \frac{1 - \sigma_1 \cdot \sigma_2}{4}, \quad P_t = \frac{3 + \sigma_1 \cdot \sigma_2}{4}. \tag{2.5}$$

We will denote

$$\frac{1}{a_s} P_s + \frac{1}{a_t} P_t = \frac{\hat{1}}{a} . \tag{2.6}$$

It is interesting to note that at the same time

$$\hat{a} = a_s P_s + a_t P_t . \tag{2.7}$$

To verify this we form the product

$$(\frac{1}{a_s} P_s + \frac{1}{a_t} P_t)(a_s P_s + a_t P_t).$$

Using the well-known properties of the projection operators P_s and P_t, it is easy to see that this product equals unity. If the system is in a definite spin state (as, for example, the singlet state of the negative hydrogen ion H^-), then one of the operators P_t or P_s (P_s in our example) will equal unity and the other zero.

Now, let us consider a system such as H + H^-. According to equations (2.4) and (2.5) we will have for the function ψ, which depends on coordinates and spin variables,

$$\frac{d}{d\rho_1} (\rho_1 \psi)_{\rho_1=0} = -[(\frac{3}{4a_t} + \frac{1}{4a_s}) + \frac{1}{4} (\frac{1}{a_t} - \frac{1}{a_s}) \sigma_1 \cdot \sigma] (\rho_1 \psi)_{\rho_1=0} \tag{2.8}$$

$$\frac{d}{d\rho_2} (\rho_2 \psi)_{\rho_2=0} = -[(\frac{3}{4a_t} + \frac{1}{4a_s}) + \frac{1}{4} (\frac{1}{a_t} - \frac{1}{a_s}) \sigma_2 \cdot \sigma] (\rho_2 \psi)_{\rho_2=0} \tag{2.9}$$

Here σ_1 and σ_2 act on the spin variables of the atoms, and σ on the spin variable of the scattered electron. ρ_1 and ρ_2 are defined by (7.4) and (7.8) of Chapter 1. In order to separate the spin variables and obtain the relations only for the coordinate part of the wave function, we must calculate the matrix elements of the spin operators $\sigma_1 \cdot \sigma$ and $\sigma_2 \cdot \sigma$.

The total spin of the system and its projection are conserved. So in the total spin representation, we need to calculate only the diagonal matrix elements.

Let us consider the state in which the total spin and its projection is 1/2. Now, there are two different ways of constructing such a state. One is to combine the electron spin state with the singlet state of both atoms; the other is to combine the electron spin state with the triplet spin state of both atoms.

The first is described by the spin wave function

$$u^{(1)} = \frac{1}{\sqrt{2}} \alpha(\alpha_1\beta_2 - \alpha_2\beta_1) \tag{2.10}$$

and the second by the function

$$u^{(2)} = \frac{1}{\sqrt{6}} [2\beta\alpha_1\alpha_2 - \alpha(\alpha_1\beta_2 + \alpha_2\beta_1)]. \tag{2.11}$$

Here α corresponds to the spin projection 1/2, and β corresponds to the spin projection $-\frac{1}{2}$. But the singlet and triplet states of a pair of hydrogen atoms have different energies. We then come to the conclusion that we have to deal with a two-channel problem.

However, the situation can be simplified if we neglect the coupling between the singlet and triplet channels of the atoms. Then the problem reduces to one-channel, and we have only to calculate the matrix element $\langle u^{(1)}|\boldsymbol{\sigma}_i\cdot\boldsymbol{\sigma}|u^{(1)}\rangle$, which is equal to 0. The final result is that, for this one channel approximation, we can use ordinary boundary conditions in which, instead of 1/a, we can substitute

$$\langle\frac{\hat{1}}{a}\rangle = \frac{1}{4}\left(\frac{3}{a_t} + \frac{1}{a_s}\right). \tag{2.12}$$

According to (7.18) of Chapter 1, the scattering amplitude is then equal to

$$A = -2\left[\frac{\cos(k\mathbf{R}\cdot\mathbf{n}/2)\cos(k\mathbf{R}\cdot\mathbf{n}_o/2)}{1/a + ik + \exp(ikR)/R} + \frac{\sin(k\mathbf{R}\cdot\mathbf{n}/2)\sin(k\mathbf{R}\cdot\mathbf{n}_o/2)}{1/a + ik - \exp(ikR)/R}\right], \tag{2.13}$$

where $n = \mathbf{r}/r$, $\mathbf{n}_o = \mathbf{k}/k$. The term $\exp(ikR)/R$ represents the influence of multiple scattering.

Note that the amplitude has a pole at some imaginary (or complex) value of k. This pole is just the energy of the particle in a bound (or quasistationary) state. If we put $k = i\alpha$ and require that

$$1/a - \alpha \pm e^{-R}/R = 0, \tag{2.14}$$

we obtain an equation for determining α as a function of R which is just the equation for the Σ_u (in the case of sign +) or Σ_g (in the case of sign -) electronic energy term in the field of two centers. Of special importance is the Σ_u term. It is responsible for a resonance in the scattering of an electron by two fixed centers.

Integration over the nuclear coordinates in (2.2) or (2.3) will smear this resonance, but nevertheless its effect will still be important.

(If we drop the multiple scattering term e^{ikR}/R from the amplitude, then we obtain for A:

$$A = -\frac{2}{1/a + ik} \cos \left[\frac{k}{2} \mathbf{R}\cdot(\mathbf{n}_0 - \mathbf{n})\right]. \tag{2.15}$$

This expression contains no resonances. The term $\cos\left[\frac{k}{2} \mathbf{R}\cdot(\mathbf{n}_0 - \mathbf{n})\right]$ represents the interference effect of two waves emitted coherently by the scattering centers.)

This simple model was used to calculate the cross section for vibrational and rotational excitation of H_2 by slow electrons (Drukarev and Yurova, 1977).

Figure 8.1 shows the differential cross section for the $0 \to 1$ vibrational transition at an energy of 3.5 eV (averaged over the initial rotational level distribution at a temperature of $T = 300^\circ K$) compared with experimental data. Figure 8.2 shows the total scattering cross section on H_2 summed over all rotational and vibrational final states compared with experimental data.

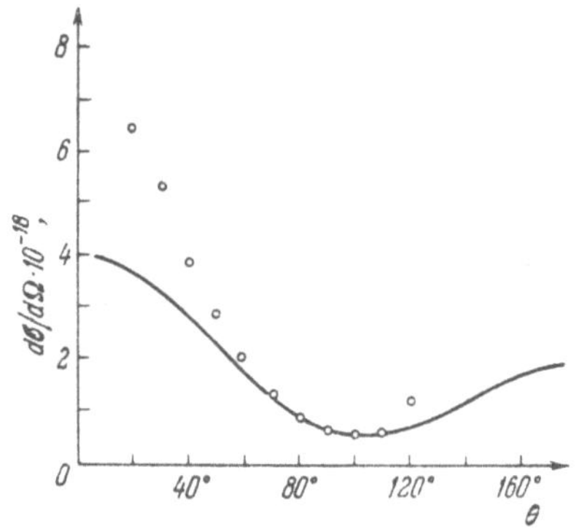

Fig. 8.1 Differential cross section for the $0 \to 1$ vibrational transition in H_2 at an incident electron energy of 3.5 eV averaged over the initial rotational level distribution at $T = 300^\circ K$.
Solid line: theory (Drukarev and Yurova, 1977).
Circles: experiment (Linder and Schmidt, 1971).
Cross section in cm^2/steradian.

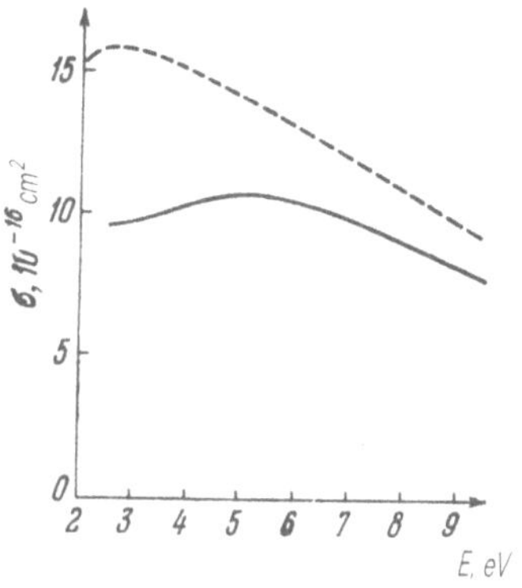

Fig. 8.2 Total scattering cross section summed over all rotational and vibrational final states.
Solid line: theory (Drukarev and Yurova, 1977).
Dashed line: experiment (Golden et al., 1966).

The adiabatic approximation outlined above is applicable in the case of a broad resonance, the corresponding lifetime being much shorter than the oscillation period of the molecule.

Now we turn to the opposite case of a long lived resonance, the oxygen molecule being an example. The energy dependence of the vibrational excitation cross section in the interval 0.3 - 1.6 eV consists of sharp peaks (Figure 8.3) located at the energies corresponding to v = 4, 5, 6 ... vibrational levels of the oxygen negative ion O_2^- (Schulz, 1973). The width of the resonances are much smaller than the distances between them. This situation is similar to the resonance scattering of slow neutrons by atomic nuclei, and we could try to apply the well-known Bohr theory of resonance reactions.

According to this theory, first the incident particle is captured by the target and a compound system is formed. Because the mean lifetime of this compound state is much longer than the characteristic time of the particle motion (in our case it is indeed the period of molecular vibration), the system "forgets" the particular way in which it was formed. As a result the decay of a compound state will be independent of its formation.

The reaction cross section then can be represented in a factorized form

$$\sigma = \sigma_c W \qquad (2.16)$$

where σ_c is the capture cross section and W is the probability of decay into a certain final state. The resonance peaks in the reaction cross sections come from the energy dependence of σ_c. If (2.16)

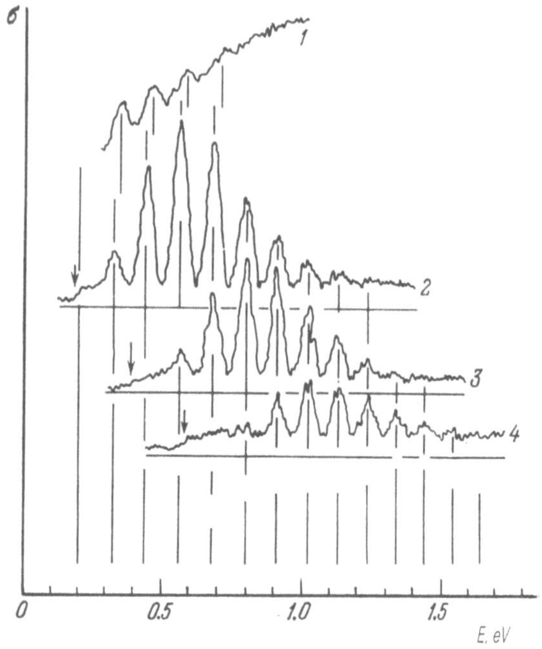

Fig. 8.3 Vibrational excitation of O_2. Measured cross sections in arbitrary units.
1. Elastic scattering; 2. v = 1 state excitation;
3. v = 2 state excitation x 2.5;
4. v = 3 state excitation x 5 (Schulz, 1973).

is correct, then the position of the resonance peaks on the energy scale should be one and the same for each particular final state.

Inspection of Figure 8.3 shows that this is the case (Schulz, 1973). One can expect also that the angular distribution of electrons after the reaction should be isotropic. Again Figure 8.4 (Schulz, 1973) confirms this conclusion (with very small deviations).

However, there are cases when neither the adiabatic approximation nor the factorization (2.16) are good approximations. As an example, vibrational excitation of the N_2 molecule can be mentioned. In Figure 8.5 the dependence of the resonance peak positions on the vibrational quantum number of the final state is shown (Schulz, 1973) in sharp contrast with Figure 8.3. Perhaps the most consistent and general treatment of vibrational excitation of a molecule by an electron is based on Fano's theory of configuration interaction (Fano, 1961).

This theory was first applied to the resonances in the photoionization of atoms. But in fact this approach can be generalized

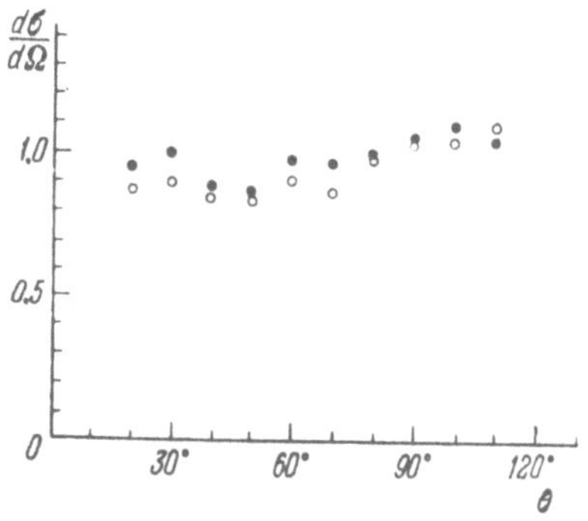

Fig. 8.4 Angular distribution of electrons scattered by O_2 molecules.
○: $v = 0 \rightarrow 1$ transition
●: $v = 0 \rightarrow 2$ transition (Schulz, 1973).

to include various kinds of resonance effects. Bardsley (1968) in his paper "Configuration interaction in continuum states of molecules" (the title of which indicates a direct and close connection to the Fano theory) derived the basic equation describing the nuclear motion in a transient negative molecular ion formed after a temporary capture of an electron by a molecule. Some improvements and further developments are contained in the paper by Fiquet-Fayard (1975) [see also the review article by Bardsley and Mandl (1968) for

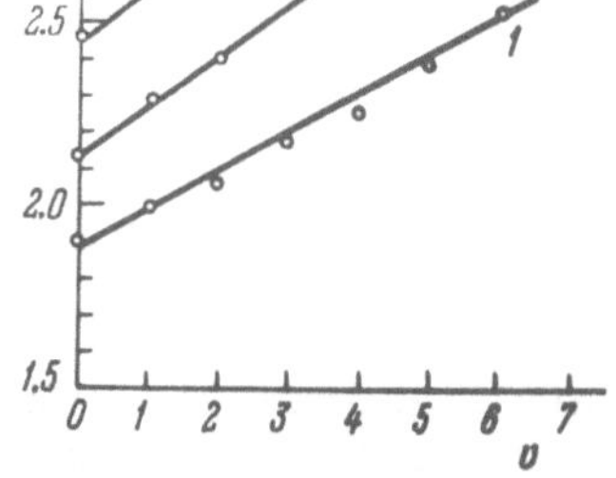

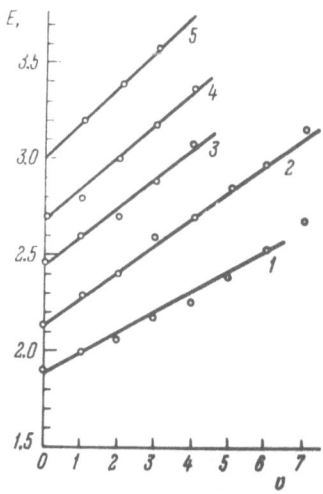

Fig. 8.5 e − N_2 scattering. Dependence of resonance peak positions upon vibrational quantum number of the final state. The peaks are identified by numbers 1 ... 5.

other references]. We outline here the main results obtained by Bardsley and Fiquet-Fayard. (However, the treatment will differ in some respects from the original papers.) To simplify the problem we assume that there is only one resonance state. Also we will neglect all interactions except that of resonance. Let us first introduce our notation. The position of the incident electron is indicated by $\mathbf{r}$, the positions of the molecular electrons are indicated by $\mathbf{r}_1 \ldots \mathbf{r}_n$. The relative position of the nuclei is indicated by $\mathbf{R}$. All electron coordinates $\mathbf{r}, \mathbf{r}_1 \ldots \mathbf{r}_n$ will be denoted for brevity by $\mathbf{q}$. The wave function of the compound state "molecule + captured electron" will be denoted by $\Phi(\mathbf{q},\mathbf{R})$ and we represent it as a product of an electronic wave function at fixed nuclei positions $\phi(\mathbf{q},\mathbf{R})$ and some function describing the nuclear motion $\xi(\mathbf{R})$ for which an equation will be derived below:

$$\Phi = \phi(\mathbf{q},\mathbf{R})\xi(\mathbf{R}). \tag{2.17}$$

The wave function of a state "vibrationally excited molecule + electron" will be denoted by $\psi_n(\mathbf{q},\mathbf{R},k)$ and expressed as a product of the molecular electronic wave function in the electronic ground state, $\eta_0(\mathbf{r}, \ldots \mathbf{r}_n,\mathbf{R})$, the nuclear wave function in n-th vibrational state, $\zeta_n(\mathbf{R})$, and the free electron wave function

$$f = Ne^{i\mathbf{k}\cdot\mathbf{r}}. \tag{2.18}$$

Introducing the unit vectors $\boldsymbol{\nu} = \mathbf{k}/k$, $\mathbf{n} = \mathbf{r}/r$, we rewrite (2.18) in the form

$$f = Ne^{ikr\boldsymbol{\nu}\cdot\mathbf{n}}. \tag{2.19}$$

We put the factor N equal to $\sqrt{\frac{1}{2\pi}}$ to ensure that f is normalized on the energy scale ($E = \frac{k^2}{2}$)

$$\int f^*(\mathbf{k}',\mathbf{r})f(\mathbf{k},\mathbf{r})d\mathbf{r} = \delta(E-E')\delta(\boldsymbol{\nu}-\boldsymbol{\nu}'). \tag{2.20}$$

Hence we have

$$\psi_n(\mathbf{q},\mathbf{R},E,\boldsymbol{\nu}) = \eta_o(\mathbf{r}_1 \ldots \mathbf{r}_n,\mathbf{R})\zeta_n(\mathbf{R})f(E,\boldsymbol{\nu},\mathbf{r}). \tag{2.21}$$

Here we neglect the exchange interaction of the scattered electron with all electrons contained in the molecule; otherwise the product $\eta_0 f$ must be replaced by a properly antisymmetrized linear combination.

Also in a more accurate treatment the plane wave (2.18) should be replaced by a continuum state electronic wave function in the field of the molecule.

Now, following Fano (1961), we express the total wave function in the form

$$\Psi = \Phi + \sum_n \int_0^\infty dE \int d\Omega \psi_n b_n(E,\nu) \tag{2.22}$$

where the coefficients $b_n(E,\nu)$ must be determined.

Some information on the analytic structure of b_n can be obtained from the condition that Ψ should have the following asymptotic form

$$\Psi_{r\to\infty} \sim \sqrt{\frac{k_o}{2\pi}}\, \eta_o \zeta_o e^{i\mathbf{k}_o\cdot\mathbf{r}} + \sum_m \frac{e^{ik_m r}}{r} \zeta_m A_m(\mathbf{n},E_m). \tag{2.23}$$

Here $\sqrt{\frac{2\pi}{k_o}}\, A_m$ is the amplitude for excitation of a vibrational state having quantum number m and E_m is the final energy of the outgoing electron, which is equal to $E_m = E_o - w_m$, w_m being the corresponding excitation energy.

Remembering the asymptotic relation

$$\int dE d\Omega \frac{e^{i\mathbf{k}\cdot\mathbf{r}}}{E - E' - i\varepsilon} F(\nu,E) = \frac{4\pi^2}{k'} \frac{e^{ik'r}}{r} F(\mathbf{n},E) \tag{2.24}$$

which holds for $r \to \infty$, $\varepsilon \to 0$, we put

$$b_n(E,\nu) = \delta(E - E_o)\delta(\nu - \nu_o) + \frac{a_n(E,\nu)}{E - E_n - i0}. \tag{2.25}$$

The coefficient a_n is closely related to the amplitude A_n:

$$A_n = \frac{4\pi^2}{\sqrt{k_n k_o}} a(E_n,\mathbf{n}). \tag{2.26}$$

Substituting (2.25) into (2.22), we get

$$\Psi = \Phi + \psi_o(E_o,\nu_o,\mathbf{R},\mathbf{q}) + \sum_n \int dE d\Omega \frac{a_n(E,\nu)\psi_n}{E - E_n - i0}. \tag{2.27}$$

Using the Schrödinger equation, we can establish a relation between a_n and the function ξ. We start from the relation

$$\int \psi^*_{n'}(E',\nu',\mathbf{R},\mathbf{q})(H - \mathcal{E})\Psi d\mathbf{R}d\mathbf{q} = 0.$$

Here H is the total Hamiltonian and $\mathcal{E}$ is the total energy which is equal to the sum of the initial energy of the molecule and of the incident electron. Inserting expression (2.27) for Ψ, taking into account the orthogonality and normalization relations

$$\int \phi \eta_o^* f^* d\mathbf{q} = 0 \tag{2.28}$$

and

$$\int \psi^*_{n'}(E',\nu',\mathbf{R},\mathbf{q})(H - \mathcal{E})\psi_n(E,\nu,\mathbf{R},\mathbf{q})d\mathbf{R}d\mathbf{q} = \delta_{nn'}\delta(\nu - \nu')\delta(E - E')(E - E_n), \tag{2.29}$$

we get

$$a_n(E,\boldsymbol{\nu}) = -\int\psi_n^* H\Phi d\mathbf{R}d\mathbf{q}. \tag{2.30}$$

Using (2.17) and (2.21) and denoting

$$\int\eta_o^* e^{-i\mathbf{k}\cdot\mathbf{r}} H\phi d\mathbf{q} = V(E,\boldsymbol{\nu},\mathbf{R}), \tag{2.31}$$

we get finally

$$a_n = -\int\xi(\mathbf{R})V(E,\boldsymbol{\nu},\mathbf{R})\zeta_n^*(\mathbf{R})d\mathbf{R}. \tag{2.32}$$

Our final step will be the derivation of an equation for the function ξ. Using the relation

$$\int\phi^*(H - \mathcal{E})\Psi d\mathbf{q} = 0, \tag{2.33}$$

inserting (2.27) and taking into account (2.32), we obtain

$$\begin{aligned}[-\frac{1}{2M} + U(R) - \mathcal{E}]\xi(\mathbf{R}) &= -\zeta_o(\mathbf{R})V(E_o,\boldsymbol{\nu}_o,\mathbf{R}) \\ &- \sum_n \zeta_n(\mathbf{R})\int \frac{dEd\Omega V^*(E,\boldsymbol{\nu},\mathbf{R})}{E - E_n - i0}\int\xi(\mathbf{R}')\zeta_n^*(\mathbf{R}')V(E,\boldsymbol{\nu},\mathbf{R}')d\mathbf{R}'.\end{aligned} \tag{2.34}$$

Using the identity

$$\frac{1}{E - E' - i0} = \mathcal{P}\frac{1}{E - E'} + i\pi\delta(E - E') \tag{2.35}$$

(here $\mathcal{P}$ means that the principal value is taken in a subsequent integration) and introducing the notation

$$\Gamma(E,\mathbf{R},\mathbf{R}') = 2\pi\int d\Omega V^*(E,\boldsymbol{\nu},\mathbf{R})V(E,\boldsymbol{\nu},\mathbf{R}'),$$

$$\Delta_o(E_n,\mathbf{R},\mathbf{R}') = \frac{1}{2\pi}\mathcal{P}\int\frac{\Gamma(E,\mathbf{R},\mathbf{R}')}{E - E_n}dE, \tag{2.36}$$

$$\mathcal{K}(\mathbf{R},\mathbf{R}') = \sum_n \zeta_n(\mathbf{R})\zeta_n^*(\mathbf{R}')[i/2\Gamma(E_n,\mathbf{R},\mathbf{R}') - \Delta_o(E_n,\mathbf{R},\mathbf{R}')],$$

we bring the equation for ξ to the final form

$$[-\frac{1}{2M}\Delta + U(R) - \mathcal{E}]\xi(\mathbf{R}) - \int\mathcal{K}(\mathbf{R},\mathbf{R}')\xi(\mathbf{R}')d\mathbf{R}' = -\zeta_o(\mathbf{R})V(E_o,\boldsymbol{\nu}_o,\mathbf{R}). \tag{2.37}$$

The asymptotic behavior of ξ as $R \to \infty$ depends on the energy $\mathcal{E}$. If it is not enough to dissociate the molecule then $\xi \to 0$ as $R \to \infty$, otherwise the asymptotic form of ξ is an outgoing wave. Now let us consider the cross section. According to (2.32) and (2.26) we have

$$\frac{d\sigma_{no}}{d\Omega} = \frac{k_n}{k_o}|A|^2 = \frac{16\pi^4}{k_o^2}|\int\xi(\mathbf{R})\zeta_n(\mathbf{R})V(E_n,\mathbf{n},\mathbf{R})d\mathbf{R}|^2. \tag{2.38}$$

The total cross section integrated over all final directions $\mathbf{n}$ and averaged over all initial directions $\boldsymbol{\nu}$ is equal to

$$\langle\sigma_n\rangle = \frac{4\pi^2}{k_0^2} \int d\mathbf{v}_o dn \left|\int \xi\zeta_n V d\mathbf{R}\right|^2. \qquad (2.39)$$

This expression can be transformed to a standard Breit-Wigner form if we introduce instead of ξ some other solution g of the same equation (2.37) which satisfies the normalization condition

$$\int |g|^2 d\mathbf{R} = 1, \qquad (2.40)$$

and we put

$$\xi = \lambda g.$$

To determine the coefficient λ we multiply both sides of (2.37) by $\lambda^* g^*$ and integrate over **R**. Introducing the notation

$$\varepsilon_o = \int g^* \left[-\frac{1}{2M}\Delta + U(R)\right] g d\mathbf{R},$$

$$\frac{\gamma}{2} = \mathrm{Im}\int g^*(\mathbf{R})\,\mathcal{K}(\mathbf{R},\mathbf{R}')g(\mathbf{R}')d\mathbf{R}d\mathbf{R}', \qquad (2.41)$$

$$\delta = \mathrm{Re}\int g^*(\mathbf{R})\,\mathcal{K}(\mathbf{R},\mathbf{R}')g(\mathbf{R}')d\mathbf{R}d\mathbf{R}',$$

we obtain the following expression for λ

$$\lambda = \frac{\int \zeta_o g^* V(E_o \mathbf{v}_o \mathbf{R}) d\mathbf{R}}{\varepsilon - \varepsilon_o + \delta - i\gamma/2}. \qquad (2.42)$$

Using this result, we get for the cross section

$$\langle\sigma_n\rangle = \frac{4\pi}{k_0^2} \frac{\int dn d\mathbf{v}_o \left|\int \zeta_o g^* V(E_o,\mathbf{v}_o,\mathbf{R})d\mathbf{R}\right|^2 \left|\int \zeta_n^* g V(E_n,\mathbf{n},\mathbf{R})d\mathbf{R}\right|^2}{(\varepsilon - \varepsilon_o + \delta)^2 + \gamma^2/4}. \qquad (2.43)$$

Now let us insert $\mathcal{K}$ from (2.36) into the expression for γ (2.41). We obtain

$$\gamma = 2\pi \sum_n \int d\mathbf{v} \left| g\zeta_n^* V(E_n,\mathbf{v},\mathbf{R})d\mathbf{R}\right|^2. \qquad (2.44)$$

Denoting by γ_n the expression

$$\gamma_n = 2\pi \int d\mathbf{v} \left|\int g\zeta_n^* V(E_n,\mathbf{v},\mathbf{R})d\mathbf{R}\right|^2, \qquad (2.45)$$

we have

$$\gamma = \sum_n \gamma_n. \qquad (2.46)$$

Finally using all the results, we reduce the expression for the cross section to the Breit-Wigner form

$$\langle\sigma_n\rangle = \frac{\pi}{k_0^2} \frac{\gamma_o\gamma_n}{(\varepsilon - \varepsilon_o + \delta)^2 + \gamma^2/4}. \qquad (2.47)$$

Here γ_n is the partial width and γ is the total width. The total cross section for excitation is equal to

$$\sum_n \langle\sigma_n\rangle = \frac{\pi}{k_0^2}\frac{\gamma_0\gamma}{(\varepsilon-\varepsilon_0+\delta)^2+\gamma^2/4}\,.$$

If the width γ is small in comparison with the average distance between the vibrational levels then the main factor which determines the energy dependence is

$$[(\varepsilon-\varepsilon_0+\delta)^2+\gamma^2/4]^{-1}.$$

The partial width γ_n will be practically constant through the resonance peak. If the width becomes larger then the oscillatory energy dependence of the partial width shows up. Let us now consider various approximations used to simplify the basic equation for ξ. If the energy of the incident electron is much larger than the distance between the vibrational levels, then it will be quite reasonable to put $E_0 \simeq E_n$ in the expression for the Kernel $\mathcal{K}(\mathbf{R},\mathbf{R}')$ in (2.36). Then using

$$\sum\zeta_n(\mathbf{R})\zeta_n^*(\mathbf{R}') = \delta(\mathbf{R}-\mathbf{R}'),$$

we get

$$\mathcal{K}(\mathbf{R},\mathbf{R}') = \delta(\mathbf{R}-\mathbf{R}')\,[\tfrac{i}{2}\,\Gamma(E_0,\mathbf{R})-\Delta_0(E_0,\mathbf{R})], \tag{2.48}$$

and the integro-differential equation (2.37) reduces to a differential equation with a complex potential

$$[-\frac{1}{2M}\Delta+U(R)+\frac{i}{2}\Gamma(E_0,\mathbf{R})-\Delta_0(E_0,\mathbf{R})-\varepsilon]\xi = -\zeta_0(\mathbf{R})V(E_0,\boldsymbol{\nu}_0,\mathbf{R}) \tag{2.49}$$

where

$$\Gamma(E,\mathbf{R}) = 2\pi\int d\boldsymbol{\nu}\,|V(E,\boldsymbol{\nu},\mathbf{R})|^2,$$
$$\Delta_0(E_0,\mathbf{R}) = \frac{1}{2\pi}\mathcal{P}\int\frac{\Gamma(E,\mathbf{R})}{E-E_0}\,dE. \tag{2.50}$$

The term $V(E_0,\boldsymbol{\nu}_0,\mathbf{R})$ which enters into the righthand side of (2.49) has a simple and clear physical meaning. It is proportional to the auto-ionization amplitude of a quasistationary negative molecular ion at fixed internuclear distance. An equation similar to (2.49) was derived using a different approach by Bardsley et al. (1966).

If we neglect the kinetic energy operator $-\frac{1}{2M}\Delta$, then we obtain the adiabatic approximation. It follows from (2.49) that in this approximation,

$$\xi = -\frac{\zeta_0(\mathbf{R})V(E_0,\boldsymbol{\nu}_0,\mathbf{R})}{(U-\Delta_0-\varepsilon)+i\,\Gamma/2}\,. \tag{2.51}$$

The approximate expression for the amplitude A_{no} will be

$$A_{no} = \frac{4\pi^2}{k} \int \zeta_n(\mathbf{R}) \frac{V}{U - \Delta_o - \varepsilon + i\,\Gamma/2} \zeta_o(\mathbf{R}) d\mathbf{R} \qquad (2.52)$$

which corresponds to the general expression (2.1), the scattering amplitude at fixed nuclei being equal to

$$A(\mathbf{R}) = \frac{V}{U - \Delta_o - \varepsilon + i\,\Gamma/2} . \qquad (2.53)$$

The resonance condition

$$\varepsilon = U + \Delta_o \qquad (2.54)$$

is up to a small shift (in comparison with $k_o^2/2$) equivalent to

$$\frac{k_o^2}{2} = U - U_o . \qquad (2.55)$$

Here $U - U_o$ is the distance between the potential curve of the negative molecular ion U and the potential curve U_o of the molecule in its initial state. This approximation is appropriate for the case of a very short lived intermediate negative molecular ion, or very large Γ. The energy dependence of the vibrational excitation cross section will have a smooth shape like the H_2 case discussed above. The expression (2.1) with A(R) given by (2.13) can be reduced to the form (2.52). In fact the physics behind (2.13) is just the same as behind (2.53), only in the former case the analysis is carried out using the multiple scattering formalism.

In the opposite case when Γ is very small, we can neglect the righthand side in (2.49) and also neglect the width on the lefthand side. Then (2.49) will reduce to

$$[- \frac{1}{2M} \Delta + U(R) - \varepsilon] \xi = 0 \qquad (2.56)$$

which is the Schrödinger equation describing the nuclear motion of the negative molecular ion as if it were stable. However, this wave function is not normalized to unity. It is the function $g = \frac{1}{\lambda} \xi$ with λ given by (2.42) which is so normalized. We then obtain the cross section in the form of the Breit-Wigner formula. The partial widths are practically constant within the narrow resonance peak. This feature is characteristic for a molecule like O_2 as was discussed above. Finally let us consider the application of equation (2.49) to N_2 and CO molecules published recently. Dube and Herzenberg (1979) and Zubec and Szmytkowski (1977) numerically integrated equation (2.49). The shape of U was approximated by the Morse potential which is customary in molecular vibration calculations. The width Γ was adjusted using the centrifugal barrier penetration factor. Rather good agreement with experiment has been achieved.

Elets and Kasansky (1981, 1982) used a semiclassical approximation to solve equation (2.49). They reduced the expression for the excitation cross section to factorized form

$$\sigma_{no} = \frac{k_n}{k_o} \frac{1}{8\pi^3} \frac{|I_1|^2 |I_2|^2}{|Q|^2} . \tag{2.57}$$

The factor $|I_1|^2$ represents the probability of capture of the incident electron into the compound negative molecular ion state. This factor is determined mainly by an integral over the product of the initial nuclear motion wave function in the target molecule and the wave function of the nuclear motion in the compound system. In fact the most important contribution to this integral comes from the region near the equilibrium nuclear position in the target molecule (region I). The factor $|I_2|^2$ represents decay of the compound system into the final vibrational-excited state of the target molecule and a free electron. It is mainly determined by an integral over the product of the final-state nuclear motion wave function in the target molecule and the wave function of the compound system. The most important contribution to the integral comes from the region near the crossing point of the compound state potential curve and the potential curve of the target molecule (region II). The factor Q represents an evolution of the compound system between the region I and II. It is equal to

$$|Q|^2 = \frac{1}{4M^2} \left| \sin \left(\int_a^b \sqrt{2M(E - W)}\, dZ - \frac{\pi}{2} \right) \right|^2 . \tag{2.58}$$

Here a and b are the "turning points," that is, the points at which the integrand is equal to zero. W is the potential energy curve of the compound state. The factors $|I_2|^2$ and $|Q|^2$ are responsible for the oscillatory behavior of the excitation cross section. The factor $|I_1|^2$ describes the shape of the envelope of these oscillations.

In Figure 8.6 the results of calculations, based on equation (2.57), for N_2 are compared with the corresponding results that were obtained by Dube and Herzenberg (1979).

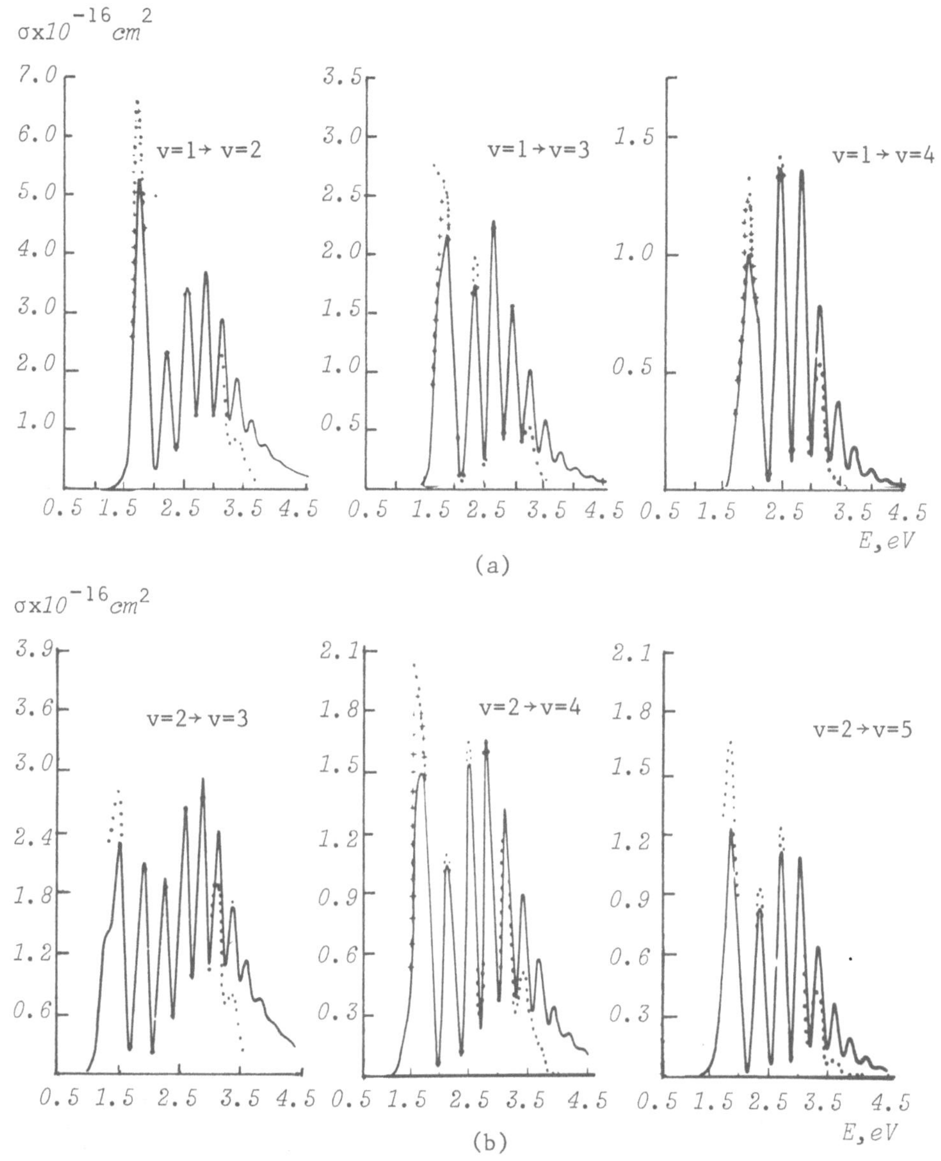

Fig. 8.6 Vibrational excitation of N_2.
a) Transitions v = 1 → 2, 1 → 3, 1 → 4.
b) Transitions v = 2 - 3, 2 - 4, 2 - 5.
Solid line: results of a numerical integration of equation (2.49) (Dube and Herzenberg, 1979).
Dotted line: semiclassical approximation (Elets and Kazansky, 1981, 1982).
Dots are shown only in the region where they deviate significantly from the solid line; in other regions the dotted line almost coincides with the solid line and is not shown.

9
Dissociation of Molecules

9.1 DISSOCIATIVE ATTACHMENT

The process of dissociative attachment

$$AB + e \rightarrow A^- + B$$

may occur whenever one of the atoms of the target molecule possesses a stable negative ion provided the energy of the incident electron exceeds the reaction threshold. Some characteristic features of this process can be demonstrated when the H_2 molecule is the target. The experimental data were summarized by Schulz (1973). In Figure 9.1 the energy dependence of the total cross section for dissociative attachment in H_2, HD and D_2 near 3.7 eV is shown. The reaction proceeds via the $^2\Sigma_u$ state of H_2^- (Figure 9.2). There is a very large isotope effect. The peak cross sections are: $1.6 \cdot 10^{-21}$ cm^2 for H_2; $1.3 \cdot 10^{-22}$ cm^2 for HD and $8 \cdot 10^{-24}$ cm^2 for D_2.

Another interesting region is near 10 eV (Figure 9.3). The reaction proceeds via the repulsive $^2\Sigma_g$ term. If we increase the electronic energy still further, then more terms come into play. Particularly near 14 eV there is a peak in the cross section due to the terms which leads to excited atomic states of the final particles (Figure 9.3). The theoretical treatment of the dissociative attachment process is based also on equation (2.49) of Chapter 8. The way in which this equation describes this process can be understood from Figure 9.4 and the following comments, reproduced from Bardsley and Mandl (1968). The nuclei are initially in some vibrational state of the target molecule. As a result of electron capture the nuclei start to move in the potential W(R) describing the compound system. Due to the repulsive nature of this potential energy, the nuclei may

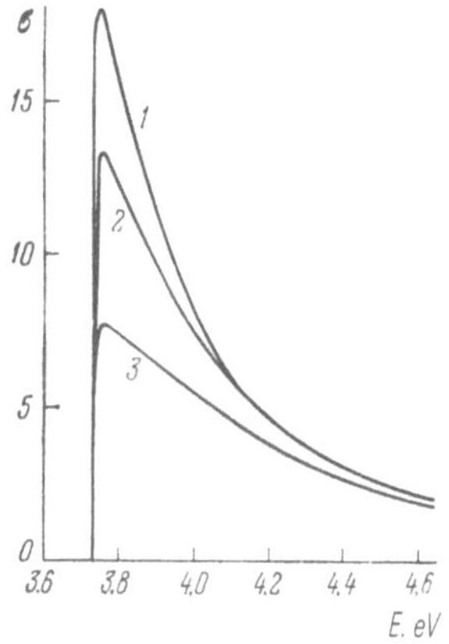

Fig. 9.1 Dissociative attachment in H_2, HD and D_2 near threshold (Schulz, 1973).
1. $(H_2) \cdot 10^{22}$ cm^2;
2. $(HD) \cdot 10^{23}$ cm^2;
3. $(D_2) \cdot 10^{24}$ cm^2.

separate to a distance R_s without auto-ionization. For R greater than the distance R_s electron emission is no longer energetically possible. The nuclear wave function ξ at large R has an asymptotic form

$$\xi \sim Ae^{ikR}/R \tag{1.1}$$

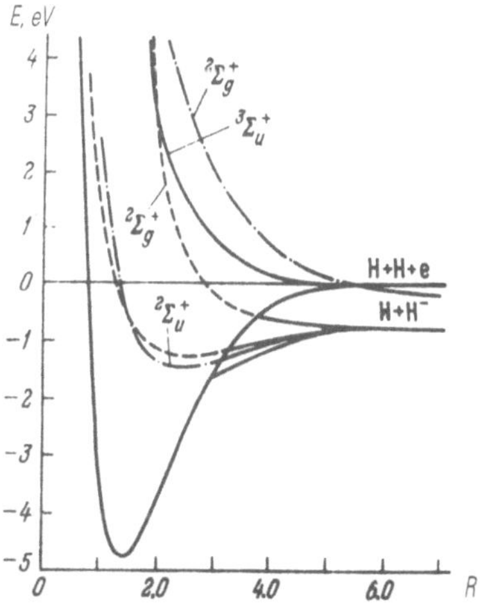

Fig. 9.2 Potential curves for H + H⁻ and H + H + e. Collection of curves obtained by different methods of calculation (Schulz, 1973).

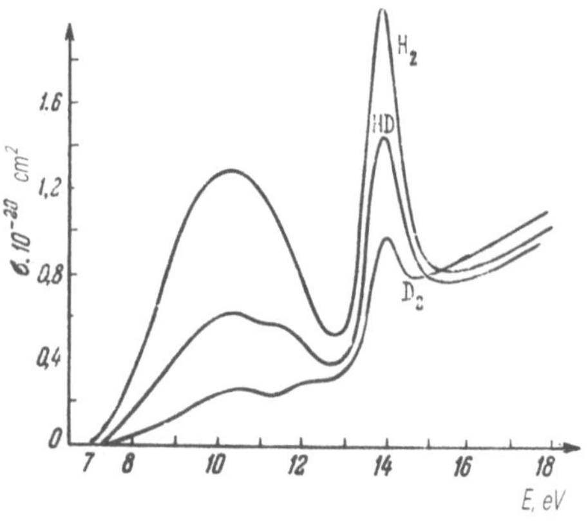

Fig. 9.3 Dissociative attachment in H_2, HD and D_2 in the 10-eV region (Schulz, 1973).

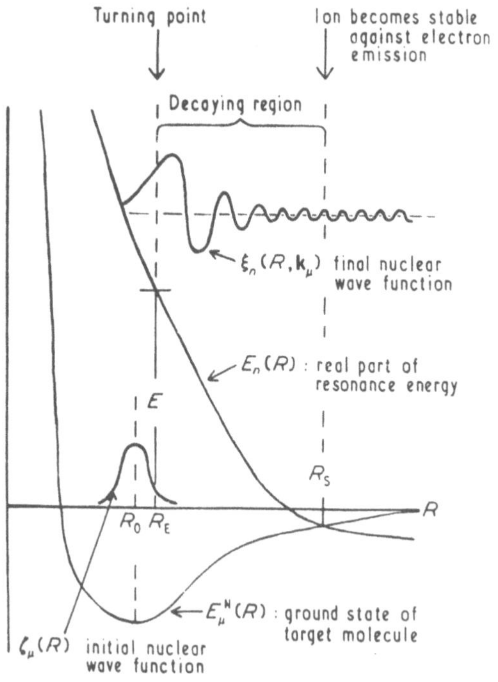

Fig. 9.4 Potential energy curves and the nuclear wave function for dissociative attachment (Bardsley and Mandl, 1968).

where A is the amplitude of the process under consideration. It follows that the cross section is equal to

$$\sigma = \frac{k}{k_o}\,|A|^2 = \frac{k}{k_o}\lim_{R\to\infty}|R\xi(R)|^2. \tag{1.2}$$

This cross section should be averaged over all directions of incident electron motion ν_o relative to the molecular axis:

$$\bar{\sigma} = \frac{k}{k_o}\,|A|^2 = \frac{k}{k_o}\frac{1}{4\pi}\lim_{R\to\infty}\int|R\xi(R)|^2 d\boldsymbol{\nu}_o. \tag{1.3}$$

In the semiclassical approximation $\bar{\sigma}$ can be expressed in the factorized form (Bardsley et al., 1966)

$$\sigma = \sigma_c \exp\left[-\int_{R_o}^{R_s}\Gamma(R)/v(R)\,dR\right] \tag{1.4}$$

where σ_c is the capture cross section and $\exp[-\int_{R_o}^{R_s}\Gamma/v dR]$ is the survival factor which expresses the probability that the nuclei reach the separation distance R_s without auto-ionization of the compound system having occurred. Because the time $1/v$ is proportional to the mass M of nuclei, the survival factor will depend strongly on M. This explains the isotope effect (Demkov, 1965). Figure 9.5 shows the results of the theoretical calculation of Bardsley et al. (1966) compared with the experimental results of Rapp et al. (1965).

Equation (2.49) of Chapter 8 was also used by Wadhera and Bardsley (1978) to investigate the effect of vibrational excitation

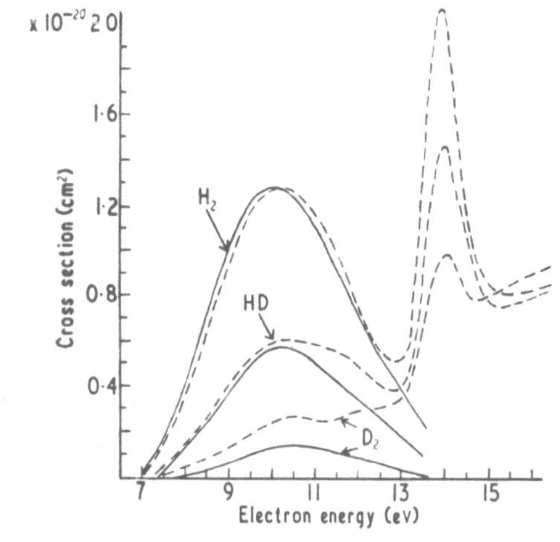

Fig. 9.5 Dissociative attachment in H_2, HD and D_2. Full curves: calculations (Bardsley et al., 1966). Broken curves: experiment (Rapp et al., 1965).

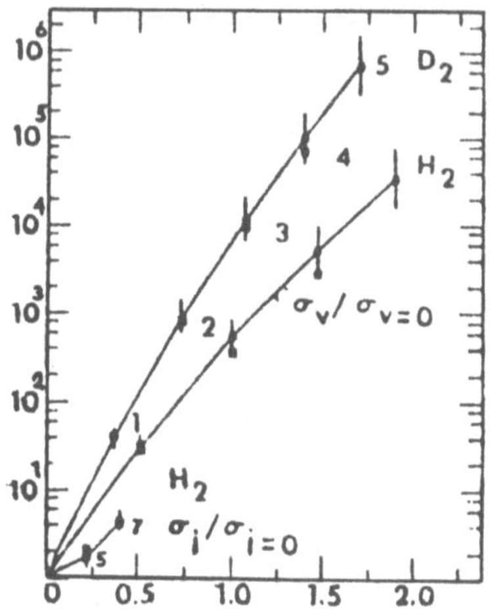

Fig. 9.6 Effect of vibrational excitation of the target molecule on the dissociative attachment cross section (Wadhera and Bardsley, 1978). The ratio of the maximum value of the dissociative attachment cross section for the excited molecule to the cross section for the molecule in the ground state is plotted against energy. Solid lines: theory; points: experiment.

of the target molecule in the initial state. This effect is very large as can be seen in Figure 9.6. Recently Hazi et al. (1981) used equation (2.37) of Chapter 8 to calculate the cross section for dissociative attachment of the F_2 molecule for the four lowest vibrational states of F_2 at electron energies from 0 to 1 eV. This is the first ab initio calculation which treats both the electronic and nuclear motions. The required electronic resonance parameters are extracted from the wave functions for $F_2(^1\Sigma_g)$ and $F_2^-(^2\Sigma_u)$ which are computed with high accuracy including target polarization and electron correlation effects. The calculated dissociation energies for $F_2(^1\Sigma_g)$ and $F_2^-(^2\Sigma_u)$ are 1.77 and 1.19 eV respectively and are in good agreement with the experimental data (1.68 and 1.29 eV). The calculated electron affinity is 2.99 eV, while the observed value is 3.4 eV. This error was corrected by shifting the potential curve of F_2^-. In Figure 9.7 the results obtained in this work are shown together with the results of other calculations and experimental data. In Figure 9.8 the cross section calculated for different initial vibrational states is shown.

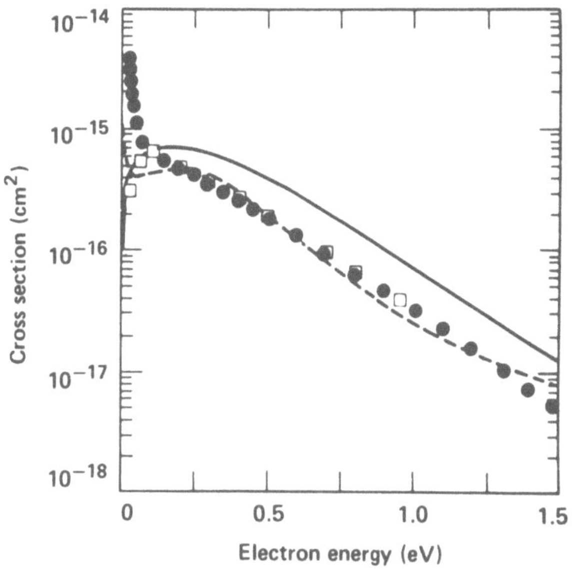

Fig. 9.7 Dissociative attachment cross section of F_2 in the ground state.
Theory: solid line, Hazi et al. (1981); dashed line, Hall (1978); squares, Bardsley et al., 1966.
Experiment: Solid circles, Chantry (private communication).

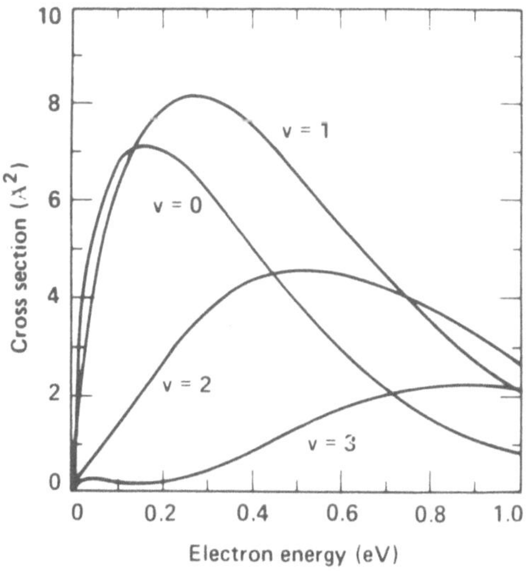

Fig. 9.8 Dissociative attachment cross section of F_2 in different initial vibration states (Hazi et al., 1981).

9.2 DISSOCIATIVE RECOMBINATION

The process of dissociative recombination

$$AB^+ + e \to A + B$$

also proceeds via the formation of an intermediate compound state $(AB)^*$. The most important energy region for the incident electron is 10^{-2} - 10^{-1} eV. There is a rather special relation of the potential energy curves of the systems $(AB)^+$, $(AB)^*$ and AB which is favorable for this process. It is shown in Figure 9.9:

a) the curve $(AB)^*$ should cross the curve AB^+ near the equilibrium point of AB^+;
b) the asymptotic position of the $(AB)^*$ curve at $R \to \infty$ should be below the ground vibrational state of $(AB)^+$.

This relation is found in H_2^+, N_2^+, NO^+, and O_2^+, and thus the cross section for these ions is large. At low energy 10^{-1} - 10^{-2} eV the expression for the survival factor in equation (1.4) is invalid. The correct expression was obtained by Bardsley (1968). We will derive this expression in a simple and straightforward way starting from equation (2.37) of Chapter 8. We will consider the case when there is no vibrational excitation. Then in the right side of (2.37) we drop all terms except the one corresponding to the ground state. We will also replace the function V(R) by its value at the equilibrium internuclear distance. Finally, we replace the plane wave describing the incident electron by a Coulomb wave function. Then for the nuclear wave function ξ, we get

$$\left[-\frac{1}{2M}\Delta + U - \varepsilon\right]\xi = \zeta_0(\mathbf{R})\{-V + [\tfrac{i}{2}\Gamma(E_0) - \Delta_0(E_0)]\int \zeta_0 \xi d\mathbf{R}\}. \quad (2.1)$$

Let us put

$$\xi = [-V + (\tfrac{i}{2}\Gamma - \Delta_0)C]f(\mathbf{R}), \quad (2.2)$$

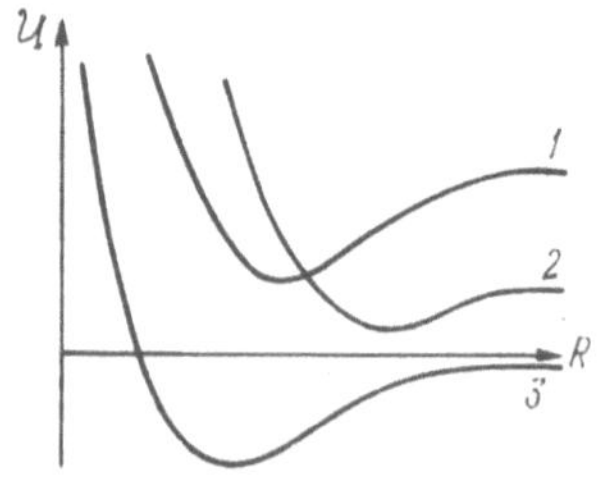

Fig. 9.9 Potential energy curves favorable for dissociative recombination.
Curve 1, $(AB)^+$; curve 2, $(AB)^*$; curve 3, AB.

where

$$C = \int \zeta_o \xi d\mathbf{R} \qquad (2.3)$$

and requires that f satisfy the equation

$$[-\frac{1}{2M}\Delta + U - \mathcal{E}]f = -\zeta_o. \qquad (2.4)$$

Inserting (2.2) into (2.3), we get for C an equation

$$C = [-V + (\frac{i\Gamma}{2} - \Delta_o)C]\int f\zeta_o d\mathbf{R}, \qquad (2.5)$$

the solution of which is

$$C = \frac{-V(E_o)\int f\zeta_o d\mathbf{R}}{[1 + \{\Delta_o(E_o) - i\Gamma(E_o)/2\}\int f\zeta_o d\mathbf{R}]}. \qquad (2.6)$$

Inserting C into (2.2), we have

$$\xi = -\frac{Vf}{1 - (i\Gamma/2 - \Delta_o)\int f\zeta_o dR}. \qquad (2.7)$$

To determine the cross section we should evaluate $\lim_{R\to\infty} |\xi R|^2$. It follows from (2.7) that

$$|\xi R|^2_{R\to\infty} = \frac{|V|^2 |fR|^2_{R\to\infty}}{[1 - \Delta_o\int f\zeta_o d\mathbf{R}]^2 + \frac{\Gamma^2}{4}|\int f\zeta_o d\mathbf{R}|^2}. \qquad (2.8)$$

Now the expression

$$S = \{|1 - \Delta_o\int f\zeta_o d\mathbf{R}|^2 + \frac{\Gamma^2}{4}|\int f\zeta_o d\mathbf{R}|^2\}^{-1} \qquad (2.9)$$

is just the desired survival factor, because it is equal to 1 when $\Gamma = 0$. The problem of the calculation of the dissociative recombination cross section reduces to the solution of (2.4). So far no attempts have been made to use equation (2.4).

9.3 DISSOCIATION THROUGH ELECTRONIC EXCITATION

We consider in conclusion the dissociation process which involves an electronic excitation of the molecule. The various types of transition are shown in Figure 9.10. Far away from threshold the adiabatic approximation can be used, according to which the amplitude is equal to

$$A_{fo} = \int \zeta_f^* \mathcal{A}_{fo}(\mathbf{R})\zeta_o d\mathbf{R}. \qquad (3.1)$$

Here $\mathcal{A}_{fo}$ is the electronic transition amplitude at fixed internuclear distance **R**, ζ_o is the ground state nuclear wave function and ζ_f is the wave function for the nuclear motion after dissociation. General

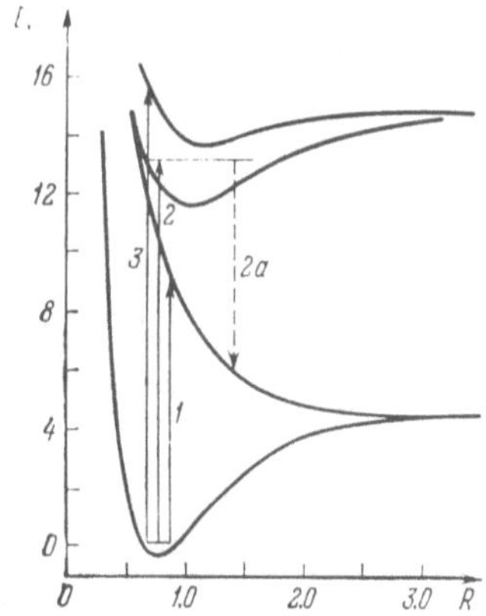

Fig. 9.10 Various types of transitions leading to dissociation of a molecule.
1. Transition to a repulsive potential energy curve.
2. Transition to an excited attractive curve followed by a radiationless transition to a repulsive one (2a).
3. Dissociative transition leading to excited atom formation.

aspects of the adiabatic approximation which involve electronic excitation were discussed by Sugard and Hazi (1975). Since the function ζ_o has a sharp maximum near the equilibrium position $\mathbf{R}_o$, it is reasonable to write approximately

$$A_{fo} = \mathcal{A}_{fo}(\mathbf{R}_o)\int \zeta_f^* \zeta_o d\mathbf{R}. \tag{3.2}$$

The integral

$$\int \zeta_f^* \zeta_o d\mathbf{R}$$

is the Franck-Condon factor. If follows from (3.2) that the cross section is equal to

$$\frac{d\sigma}{d\Omega} = \left(\frac{d\sigma}{d\Omega}\right)_{R_o} \left|\int \zeta_f^* \zeta_o d\mathbf{R}\right|^2 \tag{3.3}$$

where

$$\left(\frac{d\sigma}{d\Omega}\right)_{R_o}$$

is the electronic excitation cross section for nuclei fixed at their equilibrium position. As an example some dissociative transitions in H_2 are shown in Figure 9.11. The theoretical estimate of the dissociation cross section of the H_2 molecule was made by Cartwright

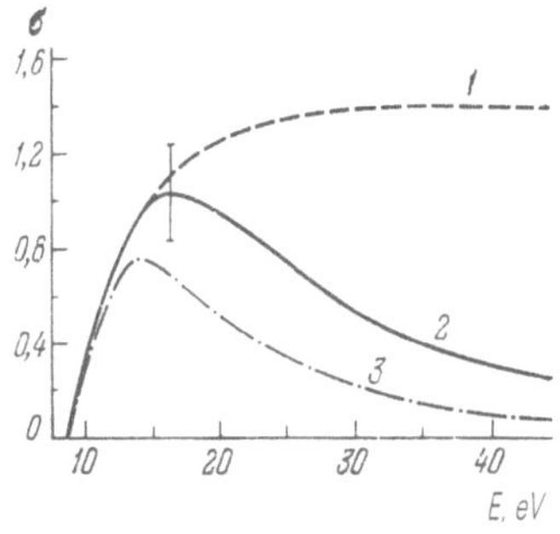

Fig. 9.11 Dissociation cross section for H_2 (Cartwright and Kuppermann, 1967).

1. Experimental data for the total dissociation cross section including the ionization of a molecule.
2. Reduced experimental curve. Contribution of ionization subtracted.
3. Calculated cross section.

and Kuppermann (1967). This estimate took into account transitions 1 and 3 shown in Figure 9.11. To calculate $(d\sigma/d\Omega)_{Ro}$ the Born-Ochkur approximation was used. The results shown in this figure are compared with experimental data.

References

Abramov, D., and Komarov, I., 1975, Vestnik Leningrad Gos. Univ., N 22: p. 24.

Abramov, D., Bemkov, Y., and Shcherbakov, A., 1981, Zh. Eksp. Teor. Fiz., 80:1334.

Abrines, R., and Percival, I. C., 1966, Proc. Phys. Soc., 88:861.

Amusia, M., et al., 1980, in: Coherence and Correlation in Atomic Physics (ed. by Kleinpoppen, H., and Williams, J.), Plenum Press.

Andrick, D., Eib, M., and Hofmann, M., 1972, J. Phys., B5:L15.

Andrick, D., 1973, Adv. Atom. Mol. Phys., 9:207.

Andrick, D., and Bitsch, A., 1975, J. Phys., B8:393.

Balashov, V., et al., 1973, Vestnik Moscow Univ., Ser. Fiz., Astron., 1:116.

Balashov, V., Grum-Grzhimailo, A. N., Kabachnik, N. M., Magunov, A. I., and Strakhova, S. I., 1980, in: Coherence and Correlation in Atomic Physics (ed. by Kleinpoppen, H., and Williams, J.), Plenum Press, p. 283.

Bardsley, J. N., Herzenberg, A., and Mandl, F., 1966, Proc. Phys. Soc., 89:305, 321.

Bardsley, J. N., and Mandl, F., 1968, Reports Progr. Phys., 31:p. 2, 471.

Bardsley, J. N., 1968, J. Phys., B1:349.

Bardsley, J. N., and Nesbet, R., 1973, Phys. Rev., A8:203.

Baz, A., 1959, Zh. Eksp. Teor. Fiz., 36:1762.

Beigman, J., and Urnov, A., 1974, J. Quant. Spectrosc. Radiat. Trans., 14:1009.

Beigman, J., Vainstein, L., and Sobelman, J., 1969, Zh. Eksp. Teor. Fiz., 57:1703.

Bethe, H. A., and Peierls, R., 1935, Proc. R. Soc. London, A148:146-156.

Blum, K., 1981, Density Matrix Theory and Applications, Plenum Press, New York, London.

Blum, K., and Kleinpoppen, H., 1979, Phys. Reports.

Bojkova, R., and Objedkov, V., 1968, Zh. Eksp. Teor. Fiz., 54:1440.

Brongersma, H. H., Knoop, F. W. E., and Backx, C., 1972, Chem. Phys. Lett., 13:16.

Brunt, J., King, G. C., and Read, F. H., 1977, J. Phys., B10:433.

Burke, P. G., Schey, H., and Smith, K., 1963, Phys. Rev., 129:1258.

Burke, P. G., McVicar, D. D., and Smith, K., 1964, Proc. Phys. Soc. (London), 83:397.

Burke, P. G., Ormonde, S., and Whitaker, W., 1967a, Proc. Phys. Soc. (London), 92:319.
Burke, P. G., Taylor, A. J., and Ormonde, S., 1967b, Proc. Phys. Soc. (London), 92:345.
Burke, P. G., Cooper, J., and Ormonde, S., 1969a, Phys. Rev., 183:245.
Burke, P. G., Gallaher, D., and Geltman, S., 1969b, J. Phys., B2:1142.
Burke, P. G., and Robb, W. D., 1975, Adv. Atom. Mol. Phys., 11:143.
Burke, P. G., 1977, Potential Scattering in Atomic Physics, Plenum Press, New York.
Burke, P. G., and Taylor, A. J., 1969, J. Phys., B2:869.
Burke, P. G., and Taylor, A. J., 1969a, J. Phys., B2:44.
Burkova, L., and Ochkur, V., 1979, Zh. Eksp. Teor. Fiz., 76:76.
Callaway, J., 1978, Physics Reports, 45:91.
Calogero, F., 1967, Variable Phase Approach to Potential Scattering, Academic Press, New York.
Cartwright, D., and Kupperman, A., 1967, Phys. Rev., 163:86.
Castillejo, L., Percival, I. C., and Seaton, M. J., 1960, Proc. Roy. Soc. London, A254:259.
Chu, S., and Dalgarno, A., 1975, Proc. R. Soc. London, A342:191.
Clementi, E., and Roetti, C., 1974, Atom. Data Nucl. Data Tables, 14:177.
Collins, R. E., Bederson, B., and Goldstein, M., 1971, Phys. Rev., A3:1976.
Cvejanovic, S., and Read, F., 1974, J. Phys., B7:1841.
Cvejanovic, S., et al., 1974a, J. Phys., B7:468.
Damburg, R., and Karule, E., 1967, Proc. Phys. Soc. (London), 90:637.
Damburg, R., and Propin, R., 1972, J. Phys., B5:533.
Demkov, Y., 1963, Variational Principles in the Theory of Collisions, Pergamon Press, London.
Demkov, Y., 1965, Phys. Lett., 15:235.
Demkov, Y., 1981, Zh. Eksp. Teor. Fiz., 80:127.
Demkov, Y., and Drukarev, G., 1965, Zh. Eksp. Teor. Fiz., 49:691.
Demkov, Y., and Ostrovsky, N., 1975, Zero Range Potential Method, Leningrad University Press.
Dolder, K. T., and Peart, K., 1973, J. Phys., B6:2415.
Drukarev, G., 1949, Zh. Eksp. Teor. Fiz., 19:247.
Drukarev, G., 1965, Theory of Collisions of Electrons with Atoms, Academic Press, New York.
Drukarev, G., and Objedkov, V., 1971, Zh. Eksp. Teor. Fiz., 61:956.
Drukarev, G., and Berezina, N., 1975, Zh. Eksp. Teor. Fiz., 69:829.
Drukarev, G., and Yurova, I., 1977, J. Phys., B10:3551.
Drukarev, G., 1978, Adv. Quant. Chem., 11:251.
Drukarev, G., Fröman, N., and Fröman, P. O., 1979, J. Phys. A12:171.
Drukarev, G., 1981, JETP., 53:271.
Dube, L., and Herzenberg, A., 1979, Phys. Rev., A20:194.
Ehrhardt, H., Jung, K., and Schubert, E., 1980, in: Coherence and Correlation in Atomic Physics (ed. by Kleinpoppen, H., and Williams, J.), Plenum Press, p. 41.
Elets, J., and Kazansky, A., 1981, Zh. Eksp. Teor. Fiz., 80:982.
Elets, J., and Kazansky, A., 1982, Zh. Eksp. Teor. Fiz., 82:450.

Elston, S., et al., 1975, Abstracts, IX ICPEAC., Seattle, p. 665.
Fabrikant, J., 1975, in: Atomic Processes, Zinatne, Riga.
Fabrikant, J., 1977, Zh. Eksp. Teor. Fiz., 73:1317
Fabrikant, J., 1978, J. Phys., B11:3621.
Fano, U., 1961, Phys. Rev., 124:1866.
Fano, U., 1957, Rev. Mod. Phys., 29:74.
Fano, U., 1974, J. Phys., B7:1401.
Fiquet-Fayard, F., 1975, J. Phys., B8:2880.
Fock, V., and Krylov, N., 1947, Zh. Eksp. Teor. Fiz., 17:93.
Foster, G., et al., 1979, Bull. Am. Phys. Soc., 24:1183.
Franco, V., 1968, Phys. Rev. Lett., 20:709.
Franco, V., 1970, Phys. Rev., A1:1705.
Franco, V., 1973, Phys. Rev., A8:2927.
Gailitis, M., and Damburg, R., 1963, Proc. Phys. Soc., London, 82:192.
Gailitis, M., 1963, Sov. Phys. JETP., 17:1328.
Gailitis, M., 1963a, Optica Spectrosc., 14:465.
Gailitis, M., 1965, in: Cross Sections of Electron-Atom Collisions (ed., Veldre, V.), p. 3, Latvian Acad. of Sci., Riga.
Gerjuoy, E., and Stein, S., 1955, Phys. Rev., 97:1761.
Gerjuoy, E., and Thomas, L., 1971, J. Math. Phys., 12:567.
Gerjuoy, E., 1971, in: Physics of Electronic and Atomic Collisions, invited papers VII ICPEAC., Amsterdam.
Giardini-Guidoni, A., Camilloni, R., and Stefani, G., 1980, in: Coherence and Correlation in Atomic Physics (ed. by Kleinpoppen, H., and Williams, J.), Plenum Press, 13.
Glauber, R., 1958, Lectures in Theoretical Physics, Boulder, USA.
Glassgold, A., and Jalongo, G., 1968, Phys. Rev., 175:151.
Golden, D., and Bandel, H., 1966, Phys. Rev., 149:58.
Golden, D., Bandel, H. W., and Salerno, J. A., 1966, Phys. Rev., 146:40.
Goldflam, R., et al., 1977, J. Chem. Phys., 67:4149, 5661.
Green and Rau, 1982, Phys. Rev. Lett., 48:533.
Hayes, M., and Seaton, M., 1978, J. Phys., B11:L79.
Hall, R. J., 1978, J. Chem. Phys., 68:1803.
Hazi, A., 1978, J. Phys., B11:1259.
Hazi, A., Orel, A., and Rescigno, T., 1981, Phys. Rev. Lett., 46:918.
Heideman, H., Nienhuis, G., and van Ittersum, T., 1974, J. Phys., B7:L493.
Heidemann, H., 1980, in: Coherence and Correlation in Atomic Physics (ed. by Kleinpoppen, H., and Williams, J.), Plenum Press.
Henry, R. J. W., 1967, Phys. Rev., 162:56.
Henry, R. J. W. et al., 1969, Phys. Rev., 178:218.
Henry, R. J. W., Burke, P. G., and Sinfailam, A. L., 1969, Phys. Rev., 178:218.
Hicks, P. J., Cvejanovic, S., Comer, J., Read, F. H., and Sharp, J. M., 1974, Vacuum, 24:573.
Inokuti, M., 1971, Rev. Mod. Phys., 43:297.
Itakava, Y., 1971, Proc. Phys. Soc., Japan, 30:835; 32:217.
John, T. L., 1960, Proc. Phys. Soc., 76:532.
Junker, B., and Huang, C., 1978, Phys. Rev., A18:313.

Junker, B., 1978, Phys. Rev., A18:2473.
Kasdan, A., Miller, T. M., and Bederson, B., 1973, Phys. Rev., A8:1562.
Karule, E., 1965, in: Cross Sections of Electron-Atom Collisions (ed., Veldre, V.), P. 3, Latvian Acad. of Sci., Riga.
Karule, E., and Peterkop, R., 1971.
Karule, E., and Peterkop, R., 1972, Akademya Nauk Latviiskoj, SSR., Izvestiya, N3, 4.
Karule, E., 1972, J. Phys., B5:2051.
Kessler, J., 1976, Polarized Electrons, Springer-Verlag, Berlin.
Khare, V., 1977, J. Chem. Phys., 68:4631.
Kingston, A. E., and Tayal, S.S., 1983, J. Phys., B16:3465.
Kleinpoppen, H., 1980, in: Coherence and Correlation in Atomic Physics (ed. by Kleinpoppen, H., and Williams, J.), Plenum Press.
Kennerly, R. E., Van Brunt, R. J., and Gallagher, A. C., 1981, Phys. Rev., A23:2430.
Krige, C., et al., 1968, Z. Naturforsch., 23a:1383.
Landau, L., and Lifschitz, E., 1974, Quantum Mechanics, 3 edn., Pergamon Press.
Lavrov, B., et al., 1980, Z. Techn. Fiz., 50:2072, 2082.
Lin, S., and Kivel, B., 1959, Phys. Rev., 114:102.
Linder, F., and Schmidt, H., 1971, Z. Naturforsch. 26a:1603, 1617.7.
Lubell, M., 1980, in: Coherence and Correlation in Atomic Physics (Ed. by Kleinpoppen, H., and Williams, J.), Plenum Press.
McCarroll, R., 1957, Proc. Phys. Soc., A70:460.
Macek, J., and Burke, P. G., 1967, Proc. Phys. Soc., 92:351.
Macek, J., and Jaecks, D. H., 1971, Phys. Rev., A4:2288
Manson, S., 1969, Phys. Rev., 192:96.
McCarthy, I., 1980, in: Coherence and Correlation in Atomic Physics (ed. by Kleinpoppen, H., and Williams, J.), Plenum Press.
McGowan, J., and Clarke, E., 1968, Phys. Rev., 167:43.
McGuire, 1971, Phys. Rev., A3:267.
Moores, D., and Norcross, D., 1972, J. Phys., B5:1482.
Morgan, L. A., 1979, J. Phys., B12:L735.
Morgenstern, R. et al., 1977, J. Phys., B10:1039.
Mott, N. F., and Massey, H. S. W., 1965, The Theory of Atomic Collisions, 3rd edn., Oxford Univ. Press.
Nesbet, R. K., 1975, Phys. Rev., A12:444.
Nesbet, R. K., 1978, J. Phys., B11:L21.
Nesbet, R. K., 1979, Phys. Rev., A20:58.
Nesbet, R. K., 1980, Variational Methods in Electron-Atom Scattering Theory, Plenum Press, New York.
Nesbet, R. K., and Lyons, J. D., 1971, Phys. Rev., A4:1812.
Neynaber, R., Marino, L. L., Rothe, E. W., and Trujillo, S. M. 1961, Phys. Rev., 123:148.
Neynaber, R., Marino, L. L., Rothe, E. W., and Trujillo, S. M., 1963, Phys. Rev., 129:2069.
Neynaber, R., 1964, in: Atomic Collision Processes, (ed. by McDowell, M. R. C., North-Holland.
Norcross, D. W., 1971, J. Phys., B4:1458.

Nussenzveig, H., 1959, Nucl. Phys., 11:499.
Oberoi, R., and Nesbet, R. K., 1973, Phys. Rev., A8:2969.
Ochkur, V., 1963, Zh. Eksp. Teor. Fiz., 45:734.
Ochkur, V., 1967, Problems of Atomic Collisions Theory, p. 1, Leningrad University Press.
Oda, N., 1973, in: Physics of Electronic and Atomic Collisions, invited papers, VIII ICPEAC., Belgrade.
O'Malley, T. F., Spruch, L., and Rosenberg, L., 1961, J. Math. Phys.
O'Malley, T. F., 1963, Phys. Rev., 130:1020.
O'Malley, T. F., Burke, P. G., and Berrington, K. A., 1979, J. Phys., B12:953.
Omidvar, K., Kyle, H. L., and Sullivan, E. C., 1972, Phys. Rev., A5:1174.
Ostrovsky, V., 1977, Sov. Phys., JETP., 72:2079.
Peach, G., 1971, J. Phys., B4:1670.
Perel, J., Englander, P., and Bederson, B., 1962, Phys. Rev., 128:1148.
Percival, I. C., and Richards, D., 1975, Adv. Atom. Mol. Phys., 11:1.
Pradhan, A., Norcross, D. W., and Hummer, D. G., 1981, Phys. Rev., A23:619.
Pradhan, A., 1981a, Phys. Rev. Lett., 47:79.
Presnyakov, L., and Urnov, A., 1975, Zh. Teor. Eksp. Fiz., 68:61.
Rapp, D., Sharp, T. E., and Briglia, D. D., 1965, Phys. Rev. Lett., 14:533.
Read, F., and Comer, J., 1980, in: Coherence and Correlation in Atomic Physics (ed. by Kleinpoppen, H., and Williams, J.), Plenum Press.
Robinson, L., 1960, Phys. Rev., 112:1281.
Roy, D., 1980, in: Coherence and Correlation in Atomic Physics (ed. by Kleinpoppen, H., and Williams, J.), Plenum Press, p. 277.
Schulz, G., 1973, Rev. Mod. Phys., 45:378.
Schwartz, C., 1961, Phys. Rev., 124:1468.
Seaton, M. J., 1958, Mon. Not. R. Astr. Soc., 118:504.
Seaton, M. J., 1969, J. Phys., B2:5.
Shimamura, I., 1979, ISAS RN83, Tokyo.
Sinfailam, A., and Nesbet, R., 1973, Phys. Rev., A7:1987.
Sinfailam, A., 1981, J. Phys., B14:L437.
Smith, A. J., Hicks, P. J., Read, F. H., Cvejanovic, S., King, G. C. M., Comer, J., and Sharp, J. M., 1974, J. Phys., B7:L496.
Spruch, L., and Rosenberg, L., 1959, Phys. Rev., 116:1034.
Spruch, L., 1967, in: The Physics of Electronic and Atomic Collisions, invited papers, V ICPEAC., Leningrad.
Stabler, R., 1964, Phys. Rev., A133:1268.
Sugard, M., and Hazi, A., 1975, Phys. Rev., A12:1895.
Sunshine, G., Aubrey, B., and Bederson, B., 1967, Phys. Rev., 154:1.
Tai, H., Bassel, R. H., Gerjuoy, E., and Franco, V., 1970, Phys. Rev., A1:1819.
Takayanagi, K., and Itikawa, Y., 1970, Adv. Atom. Mol. Phys., 6:105.
Taylor, A. J., and Burke, P. G., 1967, Proc. Phys. Soc., 92:336.

Thomas, L., Oberoi, R., and Nesbet, R., 1974, Phys. Rev., A10:1605.
Thomas, L., and Nesbet, R., 1975, Phys. Rev., A11:170; 12:1729.
Thomas, L., and Franco, V., 1976, Phys. Rev., A13:2004.
Thompson, D., 1966, Proc. R. Soc. London, A294:160.
Vainstein, L. et al., 1971, Zh. Eksp. Teor. Fiz., 61:511.
Van Blerkom, J., 1970, J. Phys., B3:932.
Varshalovich, D. et al., 1977, Astrophys. Lett., 18:167.
Veldre, V., and Vinkaln, J., 1966, in: Atomic Collisions (ed. by Veldre, V., et al.), Butterworths, London.
Vinkaln, J., and Gailitis, M., 1967, in: Abstracts of V ICPEAC., Leningrad, p. 648.
Wadhera, J., and Bardsley, J., 1978, Phys. Rev. Lett., 41:1795.
Wannier, G., 1953, Phys. Rev., 90:817.
Weigold, E., 1980, in: Coherence and Correlation in Atomic Physics (ed. by Kleinpoppen, H., and Williams, J.), Plenum Press.
Williams, J., 1974, J. Phys., B7:L56.
Williams, J., and Wills, B., 1974, J. Phys., B7:L61.
Williams, J., 1976, J. Phys., B9:1519.
Zubec, M., and Szmytkowsky, S., 1977, J. Phys., B10:L27.

Index